UNITED STATES COMMITTEE FOR THE
GLOBAL ATMOSPHERIC RESEARCH PROGRAM
National Research Council

UNDERSTANDING CLIMATIC CHANGE
A Program for Action

NATIONAL ACADEMY OF SCIENCES
WASHINGTON, D.C.
1975

WILLIAM MADISON RANDALL LIBRARY UNC AT WILMINGTON

NOTICE: The project which is the subject of this report was approved by the Governing Board of the National Research Council, acting in behalf of the National Academy of Sciences. Such approval reflects the Board's judgment that the project is of national importance and appropriate with respect to both the purposes and resources of the National Research Council.

The members of the committee selected to undertake this project and prepare this report were chosen for recognized scholarly competence and with due consideration for the balance of disciplines appropriate to the project. Responsibility for the detailed aspects of this report rests with that committee.

Each report issuing from a study committee of the National Research Council is reviewed by an independent group of qualified individuals according to procedures established and monitored by the Report Review Committee of the National Academy of Sciences. Distribution of the report is approved, by the President of the Academy, upon satisfactory completion of the review process.

The activities of the United States Committee for the Global Atmospheric Research Program leading to this report have been supported by the National Oceanic and Atmospheric Administration and the National Science Foundation under Contract NSF-C310, Task Order No. 197.

Library of Congress Cataloging in Publication Data

United States Committee for the Global Atmospheric Research Program.
   Understanding climatic change.

   Includes bibliographies.
   1. Climatic changes—Research. I. Title.
QC981.8.C5U54   1975   551.6   75-827
ISBN 0-309-02323-8

Available from

Printing and Publishing Office, National Academy of Sciences
2101 Constitution Avenue, Washington, D.C. 20418

Printed in the United States of America

U.S. COMMITTEE FOR THE GLOBAL ATMOSPHERIC RESEARCH PROGRAM

Scientific Members

Verner E. Suomi, University of Wisconsin, *Chairman*
Richard J. Reed, University of Washington, *Vice-Chairman*
Francis P. Bretherton, National Center for Atmospheric Research
T. N. Krishnamurti, Florida State University
Cecil E. Leith, National Center for Atmospheric Research
Richard S. Lindzen, Harvard University
Syukuro Manabe, Geophysical Fluid Dynamics Laboratory (NOAA)
Yale Mintz, University of California at Los Angeles
Allan R. Robinson, Harvard University
Joseph Smagorinsky, Geophysical Fluid Dynamics Laboratory (NOAA)
William L. Smith, National Environmental Satellite Service (NOAA)
Ferris Webster, Woods Hole Oceanographic Institution
Michio Yanai, University of California
John A. Young, University of Wisconsin
John S. Perry, National Research Council, *Executive Scientist*
John R. Sievers, National Research Council, *Executive Secretary*

Ex-officio Members

John W. Firor, National Center for Atmospheric Research
Robert G. Fleagle, University of Washington
Thomas F. Malone, Holcomb Research Institute

Invited Participants

Charles W. Mathews, National Aeronautics and Space Administration
Gordon H. Smith, Department of Defense
Edward P. Todd, National Science Foundation
John W. Townsend, Jr., National Oceanic and Atmospheric Administration
Robert M. White, Department of Commerce

Liaison Representatives

Eugene W. Bierly, National Science Foundation
William Chapin, Department of State
Rudolf J. Engelmann, Atomic Energy Commission
Albert Kaehn, Jr., Department of Defense
Douglas H. Sargeant, National Oceanic and Atmospheric Administration
Joseph F. Sowar, Federal Aviation Administration
Morris Tepper, National Aeronautics and Space Administration

PANEL ON CLIMATIC VARIATION

Members

W. Lawrence Gates, The Rand Corporation, *Co-Chairman*
Yale Mintz, University of California at Los Angeles, *Co-Chairman*
Wallace S. Broecker, Lamont-Doherty Geological Observatory
Kirk Bryan, Geophysical Fluid Dynamics Laboratory
Jule G. Charney, Massachusetts Institute of Technology
George H. Denton, University of Maine
Harold C. Fritts, University of Arizona
John Imbrie, Brown University
Robert Jastrow, Goddard Institute for Space Studies
Edward N. Lorenz, Massachusetts Institute of Technology
Syukuro Manabe, Geophysical Fluid Dynamics Laboratory
J. Murray Mitchell, Jr., Environmental Data Service
Jerome Namias, Scripps Institution of Oceanography
Henry Stommel, Massachusetts Institute of Technology
Warren M. Washington, National Center for Atmospheric Research

Consultants

John E. Kutzbach, University of Wisconsin
Cecil E. Leith, National Center for Atmospheric Research
Abraham H. Oort, Geophysical Fluid Dynamics Laboratory
Richard C. J. Somerville, National Center for Atmospheric Research

FOREWORD

The Preface of the U.S. Committee for the Global Atmospheric Research Program document *Plan for U.S. Participation in the Global Atmospheric Research Program* begins:

In late 1967 the International Council of Scientific Unions, acting jointly with the World Meteorological Organization, proposed a Global Atmospherc Research Program (GARP) to accomplish the objectives stated in U.N. Resolutions 1721 (XVI) and 1802 (XVII), namely, "to advance the state of atmospheric sciences and technology so as to provide greater knowledge of basic physical forces affecting climate . . .; to develop existing weather forecasting capabilities . . .," and "to develop an expanded program of atmospheric science research which will complement the program fostered by the World Meteorological Organization."

Now, in 1974, a program to reach the "weather forecasting" objective of GARP is well under way. In this report, the U.S. Committee for the Global Atmospheric Research Program (USC-GARP) outlines a program to "understand the basic physical forces affecting climate." There is ample evidence, summarized in Appendix A of this report, that climate does change, and there is more than ample evidence from past history and even recent events that changes in climate can profoundly affect human activities and even life itself. Indeed, as a growing population places ever greater demands on food and fiber resources, man's sensitivity to variations in climate will increase. We have an urgent need for better information on global climate. Unfortunately, we do not have a good quantitative understanding of our climate machine and what determines its course. Without this fundamental understanding, it does not seem possible to predict climate—neither in its short-term variations nor in its larger long-term changes. There are some who believe that important variations in climate can occur with changes in the controlling factors that are so small they are difficult to measure. With such barriers to be overcome, is there any assurance of success? We believe so.

First, the two GARP objectives, dealing with weather and climate, are strongly related to each other. A better understanding of the physical processes that affect one means a better understanding of the processes that affect the other. The difference lies mainly in how the processes should be taken into account. Mathematical models fashioned to take into account long-term changes will have to have some characteristics that are different from those fashioned mainly for short-term (weather) changes. There has been significant progress in GARP's weather objective; therefore, there already has been important progress in GARP's climate objective.

Second, there has been a tremendous improvement in our ability to observe the global weather, thanks to weather satellites. By the end of this decade, we will have the ability to observe the entire earth with needed meteorological observations. An ability to obtain better weather observations is an ability to obtain better climate observations also. Important as this is, meteorological satellites also allow us to monitor those parameters that we now believe control the climate machine: the sun's output, the earth's albedo, the distribution of clouds, the fields of ice and snow, and the temperatures of the upper layers of the ocean. These parameters control the average state of the weather and thus climate. Meteorological satellites are observing some of these parameters and could measure all of them. In some instances, the data are already being collected. These now need to be assembled to serve the needs of climate research. The feasibility of collecting data from ocean platforms has been established. A program to do it is needed. This report outlines the key requirements.

Third, research into past climates has made significant advances. We now not only know what happened in the past far better than we did a decade or two ago, but these data will provide an important information base against which theories and numerical models of climate can be tested.

Last, but far from least, there is a new generation of atmospheric scientists. Their tools are the computer, numerical models, and satellites, and they know how to use them well. The USC-GARP believes that this is an adequate manpower base. We do not expect any breakthroughs, and progress could be slower than desired, but the program outlined in this document is a rational approach toward obtaining progress as rapidly as possible on this vital subject.

The USC-GARP further believes that neither the scientific community nor the nation can afford to be complacent with its present level of understanding on this important aspect of the earth's physical environment. The natural forces determining the world's weather and climate are beyond our control, but having better insight into what nature might do should help the nation to plan for what it must do.

This report was prepared by the Committee's Panel on Climatic Variation. On behalf of the USC-GARP, I express full appreciation to its Co-Chairmen, Yale Mintz and W. Lawrence Gates, and all of its members for this important report.

VERNER E. SUOMI, *Chairman*
U.S. Committee for the
Global Atmospheric Research Program

PREFACE

The increasing realization that man's activities may be changing the climate, and mounting evidence that the earth's climates have undergone a long series of complex natural changes in the past, have brought new interest and concern to the problem of climatic variation. The importance of the problem has also been underscored by new recognition of the continuing vulnerability of man's economic and social structure to climatic variations. Our response to these concerns is the proposal of a major new program of research designed to increase our understanding of climatic change and to lay the foundation for its prediction.

The need for increased understanding of the physical basis of climate was recognized by the Panel on International Meteorological Cooperation of the Committee on Atmospheric Sciences in its report of 1966, which led to the development of the Global Atmospheric Research Program (GARP). This objective was embodied in the GARP plan as a "second objective" devoted to the study of the physical basis of climate, to be undertaken along with the program's primary concern of improving and extending weather forecasts with the aid of numerical models.

In March 1972, the United States Committee for GARP appointed the Panel on Climatic Variation to study the problem and to submit recommendations appropriate for climatic objectives of GARP observational programs, particularly the First GARP Global Experiment (FGGE) planned for 1978. The Panel's charge was subsequently enlarged to include recommendations for the design and implementation of a national climatic research program.

In its initial deliberations, the work of the Panel seemed logically to fall into three categories, depending on the time scale of climatic variation. First, the shorter-period variations, of the order 10^{-1} to 10 years, which are documented by modern instrumental observations; second, the variations of intermediate length, of the order 10 to 10^3 years, which are largely documented by historical and proxy data sources; and third, the longer-period variations, of the order 10^3 years and beyond, for which documentation comes from paleoclimatic and geological records. Three subpanels were therefore formed, and a report was issued in February 1973 by the subpanel concerned with monthly to decadal time scales (W. L. Gates, Chairman), which is the basis of the main body of the present report. The deliberations of the subpanels concerned with decadal to millenial changes (J. M. Mitchell, Chairman) and with millenial changes and beyond (W. S. Broecker, Chairman) were the basis of Appendix A of this report.

From the beginning of the Panel's work it was realized that it would be necessary to address a wide range of questions involving the use of climatic data from instrumental and proxy sources, the use of numerical simulation models, and the conduct of research on the physical mechanisms of climatic change. It was also obvious in undertaking an assignment of this magnitude that the Panel would not be able to refer to the large number of studies that have an important bearing on the problem of climatic variation. We have, therefore, generally cited only those works that were useful in framing our recommendations and in making a brief overview of present research (see Chapter 5). Some of our recommendations have been made previously by other groups [see, for example, C. L. Wilson (Chairman), 1971: Study of Man's Impact on Climate (SMIC) Report, *Inadvertent Climate Modification,* W. H. Matthews, W. W. Kellogg, and G. D. Robinson, eds. Massachusetts Institute of Technology, Cambridge, Mass.], and we are also aware that the problem of climatic change has been considered by several other groups and is of concern to other committees of GARP.

In addition to the contributions of the Panel's members, a number of consultants to the Panel also made valuable contributions: J. E. Kutzbach and A. H. Oort on the observational and statistical aspects of climatic change; C. E. Leith on the question of climatic predictability; and R. C. J. Somerville on the evaluation of numerical model performance. A. R. Robinson of Harvard University also contributed material on the role of the oceans in climatic change. Useful comments on various aspects of the Panel's work were also made by S. H. Schneider of the National Center for Atmospheric Research; by E. W. Bierly, J. O. Fletcher, and U. Radok of the National Science Foundation; by R. S.

Lindzen of Harvard University; by R. J. Reed and R. G. Fleagle of the University of Washington; and by J. Smagorinsky of the Geophysical Fluid Dynamics Laboratory. The organization and preparation of the report as a whole was undertaken by W. L. Gates.

Appendix A, which is a survey of past climatic variations, was prepared principally by J. Imbrie, W. S. Broecker, J. M. Mitchell, Jr., and J. E. Kutzbach. The portions of this Appendix concerned with dendrochronology were prepared by H. C. Fritts, and those concerned with glaciology by G. H. Denton. Unpublished data and figures used in this Appendix were also kindly supplied by A. H. Oort; C. Sancetta of Oregon State University; A. McIntyre, J. D. Hays, and G. Kukla of the Lamont-Doherty Geological Observatory; V. C. LaMarche of the University of Arizona; J. Kennett of the University of Rhode Island; and T. Kellogg, N. G. Kipp, R. K. Matthews, and T. Webb of Brown University.

Appendix B, which presents a comparative review of selected climate simulation capabilities of global general circulation models, was prepared principally by W. L. Gates, K. Bryan, and W. M. Washington. Valuable comments and contributions of unpublished material were also made by S. Manabe, R. C. J. Somerville, Y. Mintz, and R. C. Alexander of The Rand Corporation.

This report makes no claim to completeness, and many important matters are not touched upon. For example, we have not considered the questions of instrumental design and logistical support necessary to carry out the observational programs that we have recommended, nor have we dealt with the training and educational activities necessary to supply the additional scientific manpower. Although we have presented some thoughts on possible organizational arrangements for the conduct of the necessary research, and have made some preliminary cost estimates, such questions were regarded as being outside the scope of the Panel's immediate objectives and responsibility.

The principal purpose of this report is to recommend a comprehensive research program, which we feel is necessary to increase significantly our understanding of climatic variation, and the Panel will consider its efforts to have been successful if the report serves as a useful planning document to this end. In making its recommendations, the Panel is aware of what has been called the problem of "(don't know),"[2] i.e., those who are called on to implement the program may not know that we don't know the answers to the central questions. The presentation of this report at least makes it clear that *we* don't know, and thereby reduces the exponent to unity. The successful execution of the program should remove at least part of the remaining "don't know." In short, we have attempted to describe here what *should* be done, and recognize that what *can* be done and then what actually *will* be done remain to be determined.

We wish to acknowledge the valuable advice and assistance of John R. Sievers of the National Research Council and of Verner E. Suomi of the University of Wisconsin throughout the preparation of this report. We are also indebted to Viv Pickelsimer of The Rand Corporation for her efficient handling of many of the details of the Panel's work and the preparation of the typescript.

                              W. LAWRENCE GATES
                              YALE MINTZ
                              Co-Chairmen, Panel on Climatic Variation

CONTENTS

1 INTRODUCTION   1
   Limits of Our Present Knowledge / 2
      Need for Data; Need for Understanding; Need for Assessment
   Future Efforts and Resources / 5
      Research Approaches; The Question of Priorities
   Purposes and Contents of This Report / 7

2 SUMMARY OF PRINCIPAL CONCLUSIONS AND RECOMMENDATIONS   9

3 PHYSICAL BASIS OF CLIMATE AND CLIMATIC CHANGE   13
   Climatic System / 13
      Components of the System; Physical Processes of Climate; Definitions
   Causes of Climatic Change / 20
      Climatic Boundary Conditions; Climatic Change Processes and Feedback Mechanisms; Climatic Noise
   Role of the Oceans in Climatic Change / 25
      Physical Processes in the Ocean; Modeling the Oceanic Circulation
   Simulation and Predictability of Climatic Variation / 28
      Climate Modeling Problem; Predictability and the Question of Transitivity; Long-Range or Climatic Forecasting

4 PAST CLIMATIC VARIATIONS AND THE PROJECTION OF FUTURE CLIMATES — 35

Importance of Studies of Past Climates / 35
Record of Instrumentally Observed Climatic Changes / 36
Historical and Paleoclimatic Record / 37
    Nature of the Evidence; Summary of Paleoclimatic History
Inference of Future Climates from Past Behavior / 40
    Natural Climatic Variations; Man's Impact on Climate

5 SCOPE OF PRESENT RESEARCH ON CLIMATIC VARIATION — 46

Climatic Data Collection and Analysis / 46
    Atmospheric Observations; Oceanic and Other Observations; Observational Field Programs
Studies of Climate from Historical Sources / 49
Studies of Climate from Proxy Sources / 50
    General Syntheses; Chronology; Monitoring Techniques; Proxy Data Records and Their Climatic Inferences; Institutional Programs
Physical Mechanisms of Climatic Change / 54
    Physical Theories and Feedback Mechanisms; Diagnostic and Empirical Studies; Predictability and Related Theoretical Studies
Numerical Modeling of Climate and Climatic Variation / 56
    Atmospheric General Circulation Models and Related Studies; Statistical–Dynamical Models and Parameterization Studies; Oceanic General Circulation Models; Coupled General Circulation Models
Applications of Climate Models / 59
    Simulation of Past Climates; Climate Change Experiments and Sensitivity Studies; Studies of the Mutual Impacts of Climate and Man

6 A NATIONAL CLIMATIC RESEARCH PROGRAM — 62

The Approach / 63
    What Climatic Events and Processes Can We Now Identify?; Why Is a Program Necessary?
The Research Program (NCRP) / 66
    Data Needed for Climatic Research; Research Needed on Climatic Variation; Needed Applications of Climatic Studies
The Plan / 94
    Subprogram Identification; Facilities and Support; Timetable and Priorities within the Program; Administration and Coordination

A Coordinated International Climatic Research Program
(ICRP) / 105
Program Motivation and Structure; Program Elements; Program Support

REFERENCES 111

APPENDIX A: SURVEY OF PAST CLIMATES 127
Introduction / 127
Nature of Paleoclimatic Evidence; Instrumental and Historical Methods of Climate Reconstruction; Biological and Geological Methods of Climate Reconstruction; Regularities in Climatic Series
Chronology of Global Climate / 148
Period of Instrumental Observations; The Last 1000 Years; The Last 5000 Years; The Last 25,000 Years; The Last 150,000 Years; The Last 1,000,000 Years; The Last 100,000,000 Years; The Last 1,000,000,000 Years
Geographic Patterns of Climatic Change / 163
Structure Revealed by Observational Data; Structure Revealed by Paleoclimatography
Summary of the Climatic Record / 179
Future Climate: Some Inferences from Past Behavior / 182
Potential Contribution of Sinusoidal Fluctuations of Various Time Scales to the Rate of Change of Present-Day Climate; Likelihood of a Major Deterioration of Global Climate in the Years Ahead

References 190

APPENDIX B: SURVEY OF THE CLIMATE SIMULATION CAPABILITY OF GLOBAL CIRCULATION MODELS 196
Introduction / 196
Development and Uses of Numerical Modeling / 198
Atmospheric General Circulation Models / 201
Formulation; Solution Methods; Selected Climatic Simulations
Oceanic and Coupled Atmosphere–Ocean General Circulation Models / 218
Formulation; Solution Methods; Selected Climatic Simulations; Coupled Ocean–Atmosphere Models

References 236

1
INTRODUCTION

Climatic change has been a subject of intellectual interest for many years. However, there are now more compelling reasons for its study: the growing awareness that our economic and social stability is profoundly influenced by climate and that man's activities themselves may be capable of influencing the climate in possibly undesirable ways. The climates of the earth have always been changing, and they will doubtless continue to do so in the future. How large these future changes will be, and where and how rapidly they will occur, we do not know.

A major climatic change would force economic and social adjustments on a worldwide scale, because the global patterns of food production and population that have evolved are implicitly dependent on the climate of the present century. It is not primarily the advance of a major ice sheet over our farms and cities that we must fear, devastating as this would be, for such changes take thousands of years to evolve. Rather, it is persistent changes of the temperature and rainfall in areas committed to agricultural use, changes in the frost content of Canadian and Siberian soils, and changes of ocean temperature in areas of high nutrient production, for example, that are of more immediate concern. We know from experience that the world's food production is highly dependent on the occurrence of favorable weather conditions in the "breadbasket" areas during the growing seasons. Because world grain reserves are but a few percent of annual consumption, an unfavorable crop year, such as occurred in the Ukraine in 1972, has immediate international consequences. The current drought in parts of Asia and in central Africa is producing severe hardship and has already caused the migration of millions of people.

As the world's population grows and as the economic development of newer nations rises, the demand for food, water, and energy will steadily increase, while our ability to meet these needs will remain subject to the vagaries of climate. Most of the world's land suitable for agriculture or grazing has already been put to use, and many of the world's fisheries are being exploited at rates near those of natural replenishment. As we approach full utilization of the water, land, and air, which supply our food and receive our wastes, we are becoming increasingly dependent on the stability of the present seemingly "normal" climate. Our vulnerability to climatic change is seen to be all the more serious when we recognize that our present climate is in fact highly *abnormal,* and that we may already be producing climatic changes as a result of our own activities. This dependence of the nation's welfare, as well as that of the international community as a whole, should serve as a warning signal that we simply cannot afford to be unprepared for either a natural or man-made climatic catastrophe.

Reducing this climatic dependency will require coordinated management of the nation's resources on the one hand and a thorough knowledge of the climate's behavior on the other. It is therefore essential that we acquire a far greater understanding of climate and climatic change than we now possess. This knowledge will permit a rational response to climatic variations, including the systematic assessment *beforehand* of man-made influences upon the climate and will make possible an orderly economic and social adjustment to changes in climate.

LIMITS OF OUR PRESENT KNOWLEDGE

Although we have considerable knowledge of the broad characteristics of climate, we have relatively little knowledge of the major processes of climatic *change.* To acquire this knowledge it will be necessary to use all the research tools at our disposal. We must also study each component of the climatic system, which includes not only the atmosphere but the world's oceans, the ice masses, and the exposed land surface itself. Only in this way can we expect to make significant advances in our understanding of the elusive and complex processes of climatic change.

Need for Data

Observations are essential to the development of an understanding of climatic change; without them, our theories will remain theories and the potential uses of our models will remain untapped. Our observational records must be extended in both space and time, so that we can adequately document the climatic events that have occurred in the past, and so that we can monitor the climatically important physical processes that are now going on around us. Much of the present climatic data are of limited availability and need to put into forms that permit the systematic determination of appropriate climatic statistics and the assessment of the practical consequences of climatic variation. It is especially important that climatic data be organized and assembled to permit their use in conjunction with dynamical climate models.

The oceans in particular exert a powerful influence on the earth's climates, yet we have inadequate oceanographic observations on the space and time scales needed for climatic studies. The important heat, moisture, and momentum exchanges that occur at the sea surface, and the corresponding transports that occur within the ocean, are not at all well known. Recent observations from the Mid-ocean Dynamics Experiment (MODE) reveal energetic oceanic mesoscale motions at subsurface levels, and our ignorance becomes even greater than we thought it was.

The present international network of conventional meteorological observations has grown largely in response to the need for weather forecasts, while most oceanographic data have been collected from ships widely separated in space and time. For the proposed research program, these data must be supplemented by truly global observations of the large-scale geophysical boundary conditions and of the physical processes that are important in climatic change. It is here that satellite observations are expected to play a key role, as they offer an unparalleled opportunity to monitor a growing list of variables, such as cloudiness, temperature, and the extent of ice and snow. Other climatically important variables will require special monitoring programs, on either a global or regional basis. It is essential, moreover, that the relevant data be collected on a long-term basis in order to acquire the necessary statistics of climate.

Need for Understanding

Our knowledge of the mechanisms of climatic change is at least as fragmentary as our data. Not only are the basic scientific questions largely unanswered, but in many cases we do not yet know enough to pose the key questions. What are the most important causes of climatic variation, and which are the most important or most sensitive of the many processes involved in the interaction of the air, sea, ice, and land components of the climatic system? Although there is evidence of a strong coupling between the atmosphere and the ocean, for example, we cannot yet say that we understand much about its consequences for climatic change. There are also indications in paleoclimatic data that the earth's climates may be significantly influenced by the long-term astronomical variations of the sun's radiation received at the top of the atmosphere. But here again we do not yet understand the processes that may be involved.

There is no doubt that the earth's climates have changed greatly in the past and will likely change in the future. But will we be able to recognize the first phases of a truly significant climatic change when it does occur? Like the familiar events of daily weather, from which the climate is derived, climatic changes occur on a variety of space scales. These range from the change of local climate resulting from the removal of a forest, for example, to regional or global anomalies resulting from shifts of the pattern of the large-scale circulation. But unlike the weather, variations of climate take place relatively slowly, and we may think in terms of yearly, decadal, and millenial climatic changes. But the system is complex, and the search for order in the climatic record has only begun.

Even the barest outline of a theory of climate must address the key question of the predictability of climatic change. This question is closely tied to the limited predictability of the weather itself and to the predictability of the various external boundary conditions and internal transfer processes that characterize the climatic system. Although there is evidence of regularity on some time scales, the climatic record includes many seemingly irregular variations of large amplitude. How do we separate the genuine climatic signal from what may be unpredictable "noise," and to what extent are the noise and signal coupled? These are important questions, and ones to which there are no ready answers. The determination of the climate's predictability will require the further development and application of both theory and dynamical models, along with a greatly expanded data base. The answers, when they are found, will determine the limit to which we can hope to predict future climatic variations.

Special attention must be paid to the fundamental role of the world's oceans in controlling the climate. The oceans not only are the primary source of the water in the atmosphere and on the land, but they constitute a vast reservoir of thermal energy. The timing and location of the exchange of this energy with the overlying air has a profound effect on the more rapidly varying atmospheric circulation. When the dynamics of this ocean–atmosphere interaction are better known, we may find that the ocean plays a more important role than the atmosphere in climatic changes.

Need for Assessment

We should add to these limits of our present knowledge the lack of comprehensive assessment of the impacts of climatic variation on human affairs. No one doubts that there are such impacts, for the specter of drought and the consequences of persistently severe winter weather are all too familiar in many parts of the world. Even so, we must admit that we cannot now adequately answer the question: What is a change of climate worth? A farmer may know what knowledge of the climatic conditions of the next growing season would be worth to him, but the answer in terms of national and international resource planning is more elusive. This lack of assessment is brought into sharper focus when we attempt to discern the economic and social consequences of possible alternative future climates.

FUTURE EFFORTS AND RESOURCES

Research Approaches

Our future efforts must be guided by the realization that climatic changes in any one part of the world are manifestations of changes in the global climatic system. Since our fundamental goal is to increase our understanding of climatic variations to the point where we may predict (and possibly even control) them, we must subject our ideas to quantitative test wherever possible.

The recent development of satellite-based observing systems, the coming of a new generation of high-speed computers, and the emergence of models suitable for climatic simulation combine to make such an undertaking feasible at this time. The importance of climatic variations requires, moreover, that we use all methods of inquiry that are likely to yield useful information, and that we do so at the earliest possible time.

The principal approaches to the problem that are available to us are shown in Figure 1.1, and we recognize the importance of maintaining a balance of effort among them. These same approaches form the elements of the climatic research program recommended in this report and broadly cover what we believe to be the needed efforts for observation, analysis, modeling, and theory. The successful execution of the program will require contributions from the physical sciences of meteorology, oceanography, glaciology, hydrology, astronomy, geology, and paleontology and from the biological and social sciences of ecology, geography, archeology, history, economics, and sociology. A program of this sort calls for a long-term commitment from the scientific research community, from the sponsoring government agencies, and from the public.

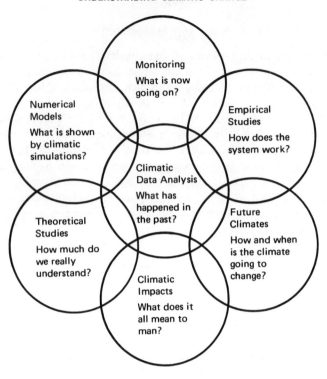

FIGURE 1.1 The interdependence of the major components of a climatic research program and a number of key questions.

The Question of Priorities

The various components of the recommended climatic research program (fully described in Chapter 6) are to a great extent interdependent: *data* are needed to check the coupled general circulation models and to calibrate the simpler models; the *models* are needed to test hypotheses and to project future climates; *monitoring* is needed to check the projections; and *all* are needed to assess the consequences. The question of priorities then becomes a matter of the priority of questions (see Figure 1.1), and there appear to be no *a priori* easy guidelines to relative importance.

Our priorities are reflected in those actions and activities that we recommend be implemented at once and in those subsequent activities

INTRODUCTION   7

for which planning should begin as soon as possible. While anticipating that much further planning will be necessary to implement the complete program, we urge that the essential interdependence of the various efforts be recognized and that all aspects of the problem be given support as parts of a coherent research program.

PURPOSES AND CONTENTS OF THIS REPORT

Broadly speaking, the purposes of this report are twofold: first, to advise the United States Government through the National Research Council's United States Committee for GARP on the urgent need for a coherent national research program on the problem of climatic variation; and, second, to advise on the steps necessary to address the same problem in the international scene.

As noted previously, our response to the Government is the recommendation of a broadly based National Climatic Research Program (NCRP), whose goal is the resolution of the problem of climatic variation. This program is presented in detail in Chapter 6, and its adoption is the first of our major national recommendations summarized in Chapter 2. In view of the possibly great impacts of future climatic variations on the nation's welfare, we believe that it is our responsibility to call for a national commitment to this effort. We accordingly urge strongly that resources to carry out such a program be made available at the earliest possible time, including provision for the necessary observations, computers, and research facilities.

Our further response to the appropriate international bodies is the proposal of a coordinated International Climatic Research Program (ICRP), which we believe to be a suitable mechanism for the pursuit of the climatic aspects of GARP. As discussed in Chapter 6, we view this as a new program of considerably greater breadth than the present GARP activities, but one for which the GARP is a necessary prelude. The U.S. national program (NCRP) would form an integral part of the ICRP, as would the national programs of other countries. In addition, we recommend a number of supporting programs whose observational requirements may impact on the First GARP Global Experiment scheduled for 1978–1979.

The remainder of this report consists of (1) a summary of our principal conclusions and recommendations (Chapter 2); (2) a discussion of the physical basis of climate and climatic change (Chapter 3); (3) a summary of past climatic variations as drawn from the instrumental and paleoclimatic record (Chapter 4); (4) a brief review of the scope of present research on climatic variation (Chapter 5); and (5) the proposed climatic research program (Chapter 6). Two technical appendixes prepared specially for this report present further details of the record and interpretation of past climates (Appendix A) and a brief comparative review of the ability of present atmospheric and oceanic general circulation models to simulate selected climatic variables (Appendix B).

2
SUMMARY OF PRINCIPAL CONCLUSIONS AND RECOMMENDATIONS

The principal conclusions and recommendations that have resulted from the deliberations of this Panel, which are expanded upon elsewhere in this report, may be summarized as follows:

1. To meet present and future national needs and to further the national contribution to GARP, we *strongly recommend* the immediate adoption and development of a coherent National Climatic Research Program (NCRP) with appropriate international coordination. The major subprograms of the NCRP are summarized in Recommendations 2, 3, and 4.

2. To perform the needed analysis of selected climatic data, including that from conventional instruments and satellites, historical records, and paleoclimatic data sources, we *recommend* the establishment of a Climatic Data Analysis Program (CDAP) as a subprogram of the NCRP. This program's functions would be to facilitate and coordinate the preparation and maintenance of a comprehensive climatic data inventory, the development of selected climatic data banks, and the preparation of suitable data analyses, based on both current and paleoclimatic data.

To carry out these functions we *recommend* the development of new climatic data-analysis facilities with access to suitable computing and data processing and display equipment, as components of a national network for climatic data analysis. We envisage these facilities as working closely with the various specialized climatic data depositories and as an essential mechanism for the successful execution of the CDAP and of related components of the overall national program.

In response to immediate practical needs, we *recommend* the initiation and continued support of empirical and statistical studies of the impacts of climatic change on man's food, water, and energy supplies. Support should also be given to studies of the broader social and economic consequences of climatic variations.

3. To acquire the needed data on the important boundary conditions and physical processes of climate, we *recommend* the development of a global Climatic Index Monitoring Program (CIMP) as a second subprogram of the NCRP. This program's functions would include the monitoring and collection, on appropriate climatic time and space scales, of data on the components of the global heat balance (including the solar constant), the ocean-surface temperature and the thermal structure of the surface mixed layer, the extent of ice and snow cover and other land-surface characteristics, the atmospheric composition and turbidity, anthropogenic processes, and, if possible, ocean-current transports and components of the hydrological cycle. This program will require a number of new observational schemes in the atmosphere, in the ocean, and on land and will rely heavily on environmental satellites. We anticipate that such data will also have important uses on a real-time basis and that the CIMP could serve as a national watchdog for climatic change.

4. To accelerate research on climatic variation, and to support the needed development of climatic modeling on a broad front, we *recommend* the establishment of a Climatic Modeling and Applications Program (CMAP) as a third subprogram of the NCRP. In this program, emphasis should be given to the development of coupled global climate models (CGCM's) of the combined atmospheric and oceanic general circulation and to the improvement of the models' treatment of clouds, mesoscale processes, and boundary-layer phenomena. Attention should also be given to the processes of air–sea interaction and to treatment of the ocean's surface layer, sea ice, and the oceanic mesoscale phenomena. We note the importance of extended model integrations to determine the annual and interannual variability of simulated climates and urge that appropriate studies be made of the sensitivity of simulated climates to physical and numerical uncertainties in the models' formulation.

To provide the basis for the needed further modeling of climatic variation, we *recommend* the development and support of a wide variety of statistical–dynamical and other parameterized climate models. We note the importance of calibration in such models and urge that appropriate schemes be developed to permit extended climatic simulations which include oceanic and cryospheric variables.

SUMMARY OF PRINCIPAL CONCLUSIONS AND RECOMMENDATIONS

To provide the needed further insight into the mechanisms of climatic variation, we *recommend* the application of climatic models in support of empirical and diagnostic studies, with particular attention to the roles of climatic feedback processes in the coupled ocean–atmosphere system, to the questions of climatic predictability and transitivity, and to the climatic effects of changes in the geophysical boundary conditions.

To provide the needed reconstruction of past climates and to develop a broader calibration of climate models, we *recommend* the initiation and support of systematic efforts to reconstruct selected events and periods in the climatic history of the earth. This should include the application of the CGCM's to simulate selected equilibrium paleoclimates and the use of statistical–dynamical or other parameterized climate models to infer the time-dependent evolution of the coupled atmosphere–hydrosphere–cryosphere climatic system.

To further the needed application of climatic models, we *recommend* the systematic exploration with suitable climate models of a variety of possible future climates, due either to natural or man-made causes. These should include determination of the likely effects of changes in solar radiation, land-surface character, cloudiness, pollution, and ice extent. We urge that efforts be made to extract consistent physical hypotheses from such experiments and that the necessary statistical controls be developed.

To lay the basis for the needed assessment of the possibilities of long-range or climatic forecasting, we further *recommend* the application of climate models of all types in experimental integrations using observed initial and boundary conditions. Appropriate climatic statistics should be drawn from such integrations and compared with observation insofar as possible, in order to establish the models' usefulness as long-range forecast tools. Initial emphasis should be given to the time periods of seasons to decades, for which there is presently the greatest practical need for scientifically based information.

To assist in the performance of the needed research or climatic modeling and applications, we *recommend* that efforts be made to identify or form a number of cooperative research associations or climatic research consortia, which we view as natural and useful coordinating mechanisms for the effective performance and long-range stability of the NCRP. We further *recommend* that the period prior to 1980 be used to develop additional scientific and technical manpower through the establishment and support of fellowships in appropriate areas of climatic research.

5. In order to further the aims of the international GARP efforts directed to the problem of climate and climatic variation, we *recommend* the adoption and development of an International Climatic Research

Program (ICRP). By the very nature of climate, the U.S. national program is considered an integral part of the ICRP, along with the climatic research programs of other nations. In view of the differences of the observational time scales and of the variables involved in weather forecasting and climatic studies, and in view of the latter's broadly interdisciplinary character, we visualize such a program being the logical successor to GARP in matters relating to climate. Recognizing that the elements of the NCRP recommended above could equally well apply to an international program, we suggest that they be considered by the appropriate international organizations.

To help provide the observational framework needed for climatic research, we *recommend* the designation of the period 1980–2000 as International Climatic Decades (ICD). During this period, efforts should be made to secure broad international cooperation in the collection, analysis, and exchange of all available climatic data, including conventional observations and special data sets of particular climatic interest (such as during droughts and following volcanic eruptions). During the ICD we also *recommend* the initiation and support of regional climatic studies in order to describe and model local climatic anomalies of special interest.

We further *recommend* development of appropriate national and international training programs and educational activities in order to promote the participation of all nations in climatic research.

6. To provide the global paleoclimatic data needed for the reconstruction of past climates, we *recommend* the development of an International Paleoclimatic Data Network (IPDN) as a subprogram of the ICRP. This program should aim to assist each nation in the cooperative identification, extraction, analysis, monitoring, and exchange of its unique paleoclimatic records, such as those from tree rings, soil types, fossil pollen, and data on sea and lake levels.

3
PHYSICAL BASIS OF CLIMATE AND CLIMATIC CHANGE

CLIMATIC SYSTEM

The term climate usually brings to mind an average regime of weather. The climatic system, however, consists of those properties and processes that are responsible for the climate and its variations and are illustrated in Figure 3.1. The properties of the climatic system may be broadly classified as thermal properties, which include the temperature of the air, water, ice, and land; kinetic properties, which include the wind and ocean currents, together with the associated vertical motions, and the motion of ice masses; aqueous properties, which include the air's moisture or humidity, the cloudiness and cloud water content, groundwater, lake levels, and the water content of snow and of land and sea ice; and static properties, which include the pressure and density of the atmosphere and ocean, the composition of the (dry) air, the oceanic salinity, and the geometric boundaries and physical constants of the system. These variables are interconnected by the various physical processes occurring in the system, such as precipitation and evaporation, radiation, and the transfer of heat and momentum by advection, convection, and turbulence.

Components of the System

In general terms the complete climatic system consists of five physical components—the atmosphere, hydrosphere, cryosphere, lithosphere, and biosphere, as follows:

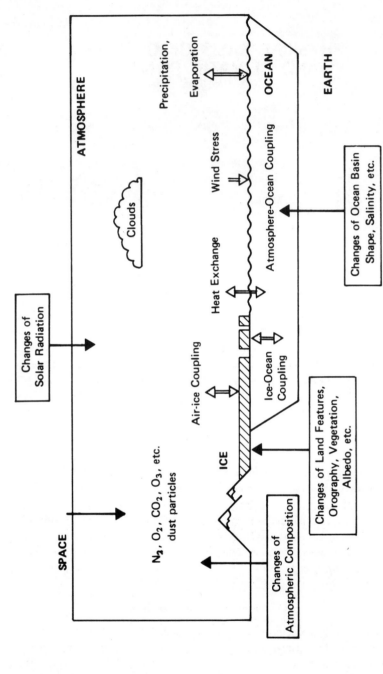

FIGURE 3.1 Schematic illustration of the components of the coupled atmosphere–ocean–ice–earth climatic system. The full arrows (⟶) are examples of external processes, and the open arrows (⟹) are examples of internal processes in climatic change.

The *atmosphere,* which comprises the earth's gaseous envelope, is the most variable part of the system and has a characteristic response or thermal adjustment time of the order of a month. By this we mean that the atmosphere, by transferring heat vertically and horizontally, will adjust itself to an imposed temperature change in about a month's time. This is also approximately the time it would take for the atmosphere's kinetic energy to be dissipated by friction, if there were no processes acting to replenish this energy.

The *hydrosphere,* which comprises the liquid water distributed over the surface of the earth, includes the oceans, lakes, rivers, and the water beneath the earth's surface, such as groundwater and subterranean water. Of these, the world's oceans are the most important for climatic variations. The ocean absorbs most of the solar radiation that reaches the earth's surface, and the oceanic temperature structure represents an enormous reservoir of energy due to the relatively large mass and specific heat of the ocean's water. The upper layers of the ocean interact with the overlying atmosphere on time scales of months to years, while the deeper ocean waters have thermal adjustment times of the order of centuries.

The *cryosphere,* which comprises the world's ice masses and snow deposits, includes the continental ice sheets, mountain glaciers, sea ice, surface snow cover, and lake and river ice. The changes of snow cover on the land are mainly seasonal and are closely tied to the atmospheric circulation. The glaciers and ice sheets (which represent the bulk of the world's freshwater storage) respond much more slowly. Because of their great mass, these systems develop a dynamics of their own, and they show significant changes in volume and extent over periods ranging from hundreds to millions of years. Such variations are, of course, closely related to the global hydrologic balance and to variations of sea level (see Appendix A).

The *lithosphere,* which consists of the land masses over the surface of the earth, includes the mountains and ocean basins, together with the surface rock, sediments, and soil. These features change over the longest time scales of all the components of the climatic system, ranging up to the age of the earth itself. The processes of continental drift and sea-floor spreading, which have resulted in mountain building and in changes in the shapes and depths of the oceans, occur over tens and hundreds of millions of years. These events are not generally regarded as representing the same kind of interaction with other components of the system as the variations described above. We note, however, that there may be a significant relationship between the occurrence of major glacial periods and the times when continental land masses occupied positions near the rotational poles of the earth (see Appendix A). The processes of isostatic adjustment and the accumulation of deep-ocean sediments also represent significant changes of the lithosphere, and as such may be viewed as earth–ice–ocean interactions. The introduction of volcanic debris into the atmosphere and its subsequent dispersal may also be cited as an example of earth–air interaction.

The *biosphere* includes the plant cover on land and in the ocean and the animals of the air, sea, and land, including man himself. Although their response characteristics differ widely, these biological elements are sensitive to climate and, in turn, may influence climatic changes. It is from the biosphere that we obtain most of the data on paleoclimates (see Appendix A). Natural changes in surface vegetation occur over periods ranging from decades to thousands of years in response to changes in temperature and precipitation and, in turn, alter the surface albedo and roughness, evaporation, and ground hydrology. Changes in animal populations also reflect climatic variations through the availability of suitable food and habitat. The anthropogenic changes due to agriculture and animal husbandry are not known but may well be appreciable in altering at least regional climates.

Physical Processes of Climate

The climate at any particular time represents in some sense the average of the various elements of weather, along with the state of the other components of the system. The physical processes responsible for climate (as distinct from climatic change) are therefore basically the same as those responsible for weather. These processes are expressed in quantitative fashion by the dynamical equation of motion, the thermodynamic energy equation, and the equations of mass and water substance continuity, as applied to the atmosphere and ocean (see Appendix B).

A process of primary importance for the circulation of the atmosphere and ocean is the rate at which heat is added to the system, the ultimate source of which is the sun's radiation. The atmosphere and ocean respond to this heating by developing winds and currents, which serve to transport heat from regions where it is received in abundance, such as in the equatorial and tropical areas, to regions where relatively little radiation is received, such as the polar regions of the earth. In this way, the atmosphere and ocean maintain the overall global balance of heat. A great deal of this heat is transported by the disturbances responsible for much of our weather in middle and high latitudes, and similar disturbances may occur in the ocean. These eddies of the general circulation also participate in the transports necessary to maintain the global balances of momentum, mass, and the total quantity of water substance.

While this simple view is a fair summary of our basic understanding of the general circulation, it is not without shortcomings. For example, it does not consider the basically different circulation regime in the low latitudes or the role of convective phenomena, and it does not consider the important variations of the circulation with height. It might also be noted that for other combinations of the planetary size and rotation rate, atmospheric composition, and meridional heating gradient, such as occur on other planets, an altogether different circulation regime—and hence climate—could result.

Although the equations referred to above are fundamental in that they form the basis of our ability to simulate numerically the climate with dynamical models, they are not in themselves particularly revealing as far as the more subtle physical processes of climate are concerned, to say nothing of the processes of climatic change. The heating rate is itself highly dependent on the distribution of the temperature and moisture in the atmosphere and owes much to the release of the latent heat of condensation during the formation of clouds and to the subsequent influence of the clouds on the solar and terrestrial radiation. These processes, together with others that contribute to the overall heat balance of the atmosphere, are shown in Figure 3.2, in which data derived from recent satellite observations have been incorporated (see, for example, Vonder Haar and Suomi, 1971). Here the presence of clouds, water vapor, and CO_2 is seen to account for over 90 percent of the long-wave radiation leaving the earth–ocean–atmosphere system. This effective blocking of the radiation emitted by the earth's surface, commonly referred to as the greenhouse effect, permits a somewhat higher surface temperature than would otherwise be the case. It is interesting that this important effect is achieved by gases in the atmosphere that exist in near trace amounts.

We see from Figure 3.2 that the role played by clouds is an important one: the reflection and emission from clouds accounts for about 46 percent of the total radiation leaving the atmosphere; and in terms of the shortwave radiation alone, clouds account for two thirds of the planetary albedo. The largest single heat source for the atmosphere is that supplied by the release of the latent heat of condensation, and this is particularly important in the lower latitudes. There is also an appreciable supply of sensible heat from the oceans, especially in the middle and higher latitudes. It is therefore clear that water substance, in either vapor or droplet form, plays a dominant role in the atmospheric heat balance. And when we recall that the oceans themselves absorb

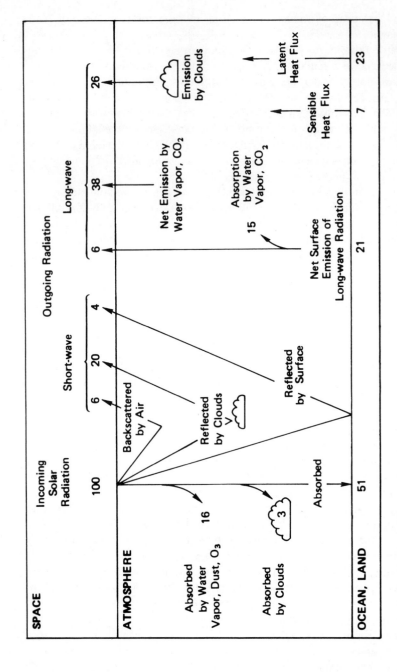

FIGURE 3.2 The mean annual radiation and heat balance of the atmosphere, relative to 100 units of incoming solar radiation, based on satellite measurements and conventional observations.

most of the solar radiation reaching the surface, and that the presence of ice and snow also affect the heat balance, the climatic dominance of global water substance becomes overwhelming, even if ice is not taken into account.

Definitions

It is useful at this point to introduce a number of definitions related to climate and climatic change. In what may be called the "common" definition, climate is the average of the various weather elements, usually taken over a particular 30-year period. A more useful definition is what we shall call the "practical" definition, which introduces the concept of a climatic state. This and related definitions are as follows:

Climatic state. This is defined as the average (together with the variability and other statistics) of the complete set of atmospheric, hydrospheric, and cryospheric variables over a specified period of time in a specified domain of the earth–atmosphere system. The time interval is understood to be considerably longer than the life span of individual synoptic weather systems (of the order of several days) and longer than the theoretical time limit over which the behavior of the atmosphere can be locally predicted (of the order of several weeks). We may thus speak, for example, of monthly, seasonal, yearly, or decadal climatic states.

Climatic variation. This is defined as the difference between climatic states of the same kind, as between two Januaries or between two decades. We may thus speak, for example, of monthly, seasonal, yearly, or decadal climatic variations in a precise way. The phrase "climatic change" is used in a more general fashion but is generally synonymous with this definition.

Climatic anomaly. This we define as the deviation of a particular climatic state from the average of a (relatively) large number of climatic states of the same kind. We may thus speak, for example, of the climatic anomaly represented by a particular January or by a particular year.

Climatic variability. This we define as the variance among a number of climatic states of the same kind. We may thus speak, for example, of monthly, seasonal, yearly, or decadal climatic variability. Although it may be confusing, this definition of climatic variability includes the variance of the variability of the individual climatic states.

The foregoing definitions are useful for two reasons. First, the concept of climatic state preserves the essence of what is usually connoted by climate, while circumventing troublesome problems of statistical stability. Second, climatic states represent definite realizations or samples of climate (rather than the climate *per se*) and are comparable with the climates simulated by numerical general circulation experiments. There are many other definitions in existence to distinguish particular statistical characteristics of climate and climatic change (such as climatic fluctuations, oscillations, periods, cycles, trends, and rhythms). The above definitions are generally adequate for our purposes, although we shall later consider another definition of climate related to the climatic system. We shall also subsequently introduce the concepts of climatic noise and climatic predictability. Except when otherwise indicated, the use of the word "climate" in this report is to be considered an abbreviation for climatic state.

It should be noted that we have included the oceans in the definition of a climatic state, as well as information on other aspects of the physical environment. The ensemble of statistics required to completely describe a climatic state is presently available for only a few regions and for limited periods of time. The climatic data-analysis and monitoring programs recommended in Chapter 6 are intended to fill in as much of the gap as possible with available data and to ensure that at least certain critical data are systematically gathered for an extended period of time in the future.

CAUSES OF CLIMATIC CHANGE

While the above discussion may describe the processes responsible for the maintenance of climate, it is an inadequate description of the processes involved in climatic *change*. Here we are on less secure ground and must consider a wide range of possible interactions among the elements of the climatic system. It is these interactions that are responsible for the complexity of climatic variation.

Climatic Boundary Conditions

If we view the gaseous, liquid, and ice envelopes surrounding the earth as the internal climatic system, we may regard the underlying ground and the space surrounding the earth as the external system. The boundary conditions then consist of the configuration of the earth's crust and the state of the sun itself. Changes in these conditions can obviously alter the state of the climatic system, i.e., they can be causes of climatic variation.

Each of the external processes illustrated in Figure 3.1 may be used to develop a climatic theory, on which basis one may attempt to explain

PHYSICAL BASIS OF CLIMATE AND CLIMATIC CHANGE 21

certain features of the observed climatic changes. For example, changes of the distribution of solar radiation have been used since the time of Milankovitch (1930) to explain the major glacial–interglacial cycles of the order of 10^4 to 10^5 years. Aside from the question of variations of the sun's radiative output, variations of the earth's orbital parameters produce changes in the intensity and geographical pattern of the seasonal and annual radiation received at the top of the atmosphere and in the length of the radiational seasons in each hemisphere. These effects, which are known with considerable accuracy, have resulted in occasional variations of the seasonal insolation regime several times larger than those now experienced. These orbital elements (eccentricity, obliquity, and precession) vary with periods averaging about 96,000 years, 41,000 years, and 21,000 years, respectively. Because the seasons themselves represent substantial climatic variations, such astronomical theories of climatic change must be given careful consideration.

The separate question of the climatic effects of possible changes in the sun's radiation (i.e., changes of the so-called solar constant) has a much less firm physical basis. Not only are the measured short-period variations of solar output quite small, but the repeated search for climatic periodicities linked with the 11-year and 80-year sunspot cycles has not yielded statistically conclusive results. The question of still longer-period solar variations cannot be adequately examined with present data, although over periods of the order of hundreds of millions of years the sun's radiation seems likely to have changed. The time range of this and other possible causative factors of climatic change is shown in Figure 3.3.

On time scales of tens of millions of years there are changes in the shapes of the ocean basins and the distribution of continents as a result of sea-floor spreading and continental drift (see Figure 3.3). Over geological time, these processes must have resulted in substantial changes of global climate. Just how much of the recorded paleoclimatic variations may eventually be accounted for by such effects, however, is not known, and applying climatic models to the systematic reconstruction of the earth's climatic history prior to about 10 million years ago is an important component of the research program recommended in this report (see Chapter 6). In such climatic reconstructions, the oceans must be simulated along with the atmosphere, and eventually the ice masses must also be reproduced. Accompanying the migration of the land masses are the processes of mountain building, epeirogeny, isostatic adjustment, and sea-level changes, all of which must also be taken into account.

Yet another external cause of climatic variation is the changes in

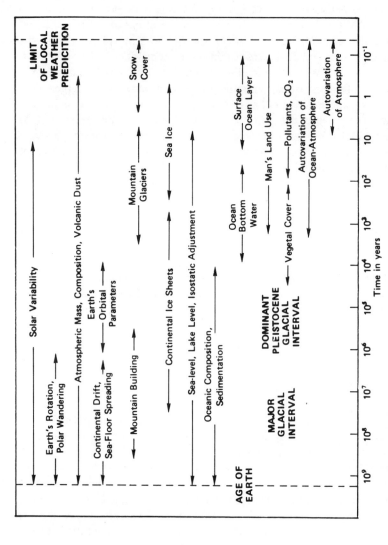

FIGURE 3.3 Characteristic climatic events and processes in the atmosphere, hydrosphere, cryosphere, lithosphere, and biosphere and possible causative factors of global climatic change.

the composition of the atmosphere resulting from the natural chemical evolution of the nitrogen, oxygen, and carbon dioxide content in response to geological and biological processes, as well as from the effluents of volcanic eruptions. On shorter time scales, however, it is probably the injection of dust particles into the atmosphere by volcanoes that has produced a more significant climatic effect by modifying the atmospheric radiation balance (see Figure 3.2). The progressive enrichment of the atmospheric CO_2 content, which has occurred during this century as a result of man's combustion of fossil fuels (amounting to an increase of order 10 percent since the 1880's), must also be considered an external cause of climatic variation.

These considerations lead to the "physical" definition of climate as the equilibrium statistical state reached by the elements of the atmosphere, hydrosphere, and cryosphere under a set of given and fixed external boundary conditions. There is, of course, the possibility that a true equilibrium may not be reached in a finite time due to the disparity of the response times of the system's components, but this is nevertheless a useful definition. By progressively reducing the internal climatic system to include only the atmosphere and ocean (in equilibrium with the land and ice distribution), and then to include only the atmosphere itself (in equilibrium with the ocean, ice, and land), a hierarchy of climates may be defined which is useful for the analysis of questions of climatic determinism.

Climatic Change Processes and Feedback Mechanisms

Important as the above processes may be for the longer-period variations of climate, there are other factors that may also produce climatic change. These involve changes in the large-scale distribution of the effective internal driving mechanisms for the atmosphere and ocean.

Variations of the global ice distribution, for example, have a significant effect on the net heating of the atmosphere (by virtue of the ice's effective control of the surface heat budget), and thereby may change the meridional heating gradient that drives the atmospheric (and oceanic) circulation. An equally significant change (for the oceans, at least) may be introduced by widespread salinity variations, as caused, for example, by the melting of ice. The salinity of the ocean surface water is in turn closely related to the formation of relatively dense bottom water, which by sinking and spreading fills the bulk of the world's ocean basins.

Such processes may act as internal controls of the climatic system, with time scales extending from fractions of a year to hundreds and even thousands of years (see Figure 3.3). Some of these processes display a coupling or mutual compensation among two (or more) elements of the internal climatic system. Such interactions or feedback mechanisms may act either to amplify the value or anomaly of one of the interacting elements (positive feedback) or to damp it (negative feedback). Because of the large number of degrees of freedom of the ocean–atmosphere system (for the moment considering the ice distribution to be fixed), there are a large number of possible feedback mechanisms within the ocean, within the atmosphere, and *between* the ocean and the atmosphere. These same degrees of freedom, however, invite a high risk of error in any qualitative analysis, and in some cases equally plausible arguments of this sort lead to opposite conclusions.

Some of the more prominent feedback effects operate among the shorter-period processes of climatic change, especially those concerning the radiation balance over land and the energy balance over the ocean. For example, a perturbation of the ocean-surface temperature may modify the transfer of sensible heat to the overlying atmosphere, and thereby affect the atmospheric circulation and cloudiness. These changes may in turn affect the ocean-surface temperatures through changes in radiation, wind-induced mixing, advection, and convergence (and may subsequently affect the deep-ocean temperatures through geostrophic adjustment to the convergence in the boundary layer). These processes may result either in the enhancement or reduction of the initial anomaly of sea-surface temperature. A number of studies have shown positive feedback of this sort for several years' time in the North Pacific Ocean.

The greenhouse effect (in which the absorption of long-wave radiation by water vapor produces a higher surface temperature), is probably the best known example of a semipermanent positive feedback process, although other positive feedbacks of climatic importance may be noted. One of these is the snow cover–albedo–temperature feedback, in which an increase of snow extent increases the surface albedo and thereby lowers the surface temperature. This in turn (all else being equal) further increases the extent of the snow cover. An example of negative feedback is the coupling between cloudiness and surface temperature noted earlier. In this scheme, an initial increase of surface temperature serves to increase the evaporation, which is followed by an increase of cloudiness. This in turn reduces the solar radiation reaching the surface and thereby lowers the initial temperature anomaly. Here we have ignored the effects of long-wave radiation and of advective processes in both ocean and atmosphere, but these examples serve to illustrate the uncertainty that must be attached to such arguments.

While there is much evidence to support the existence of feedback processes, the key phrase in their qualitative use is "all else being equal." In a system as complex as climate, this is usually not the case, and an anomaly in one part of the system may be expected to set off a whole series of adjustments, depending on the type, location, and magnitude of the disturbance. Any positive feedback must, in any event, be checked at some level by the intervention of other internal adjustment processes, or the climate would exhibit a runaway behavior. We do not adequately understand these adjustment mechanisms, and their systematic *quantitative* exploration by numerical climate models is an important task for the future (see Chapter 6). In that research it will be essential to use *coupled* models of atmosphere and ocean, and these must be calibrated with great care so as not to distort the feedback mechanisms themselves.

Climatic Noise

Climatic states have been defined in terms of finite time averages and as such are subject to fluctuations of statistical origin in addition to the changes of a physical nature already discussed. Since these statistical fluctuations arise from the day-to-day fluctuations in weather (the autovariation of the atmosphere identified in Figure 3.3), they are unpredictable over time scales of climatological interest and are therefore appropriately defined as "climatic noise." The amplitude of this noise decreases approximately as the square root of the length of the time-averaging interval, but some remains at any finite time scale (Leith, 1973; Chervin et al., 1974). A key problem of climatic variation on any time scale is therefore the determination of the "climatic predictability," which we may define as the ratio of the magnitude of the potentially predictable climatic change of physical origin to the magnitude of this unpredictable climatic noise.

ROLE OF THE OCEANS IN CLIMATIC CHANGE

It has been noted that the oceans play a prominent role in the determination of climate through the processes at the air–sea interface that govern the exchanges of heat, moisture, and momentum. While these conditions are actually determined mutually by the atmosphere and the ocean, they are likely dominated by the ocean on at least the longer climatic time scales. It is the high thermal and mechanical oceanic inertia that requires that special consideration be given to the role of the ocean in climatic change.

Physical Processes in the Ocean

Over half of the solar radiation reaching the earth's surface is absorbed by the sea. This solar radiation, along with the surface wind stress, is the ultimate energy source for a variety of physical processes in the ocean whose climatic importance is essentially a function of their time scales. The absorption of solar radiation is primarily responsible for the existence of a warm surface mixed layer of order 10^2 m deep found over most of the world's oceans. This warm surface layer represents a large reservoir of heat and acts as a significant thermodynamic constraint on the atmospheric circulation.

The exchange of the ocean's heat with the atmosphere occurs over a wide range of time scales and largely determines the relative importance of other physical processes in the ocean for climatic change. Some of this heat is used for surface evaporation, some is stored in the surface layer, and some is moved downward into deeper water by various dynamical and thermodynamical processes. The fluxes of latent and sensible heat into the atmosphere are commonly parameterized in atmospheric models as functions of the large-scale surface wind speed and the vertical gradients of humidity and temperature in the air over the ocean surface. These fluxes are actually accomplished by small-scale turbulent processes in the surface boundary layer whose behavior is not adequately understood. Physical processes in the ocean such as vertical convective motions (depending on the local vertical stratification of temperature and salinity) and wind-induced stirring also affect the depth and structure of the mixed layer, as shown, for example, by the simulations of daily variations of local mixed layer depth by Denman and Miyake (1973). Other small-scale processes such as salt fingering and internal waves also produce transports that may contribute significantly to the overall vertical mixing in the ocean. Therefore, the dynamics of the ocean's surface layer must be taken into account in even the simplest of climate models.

It is becoming apparent that the most energetic motion scale in the oceans is that of the mesoscale eddy, whose period is of the order of a few months and whose horizontal wavelength is of the order of several hundred kilometers. The kinetic energy of these motions, which is predominantly in the barotropic and first baroclinic vertical mode, may be one or two orders of magnitude greater than that of the time-averaged motions themselves. In a general sense, these slowly evolving eddies are the counterpart of the larger-scale transient cyclones and anticyclones in the atmosphere. An understanding of the physical processes responsible for the origin and behavior of these eddies and their role in the oceanic general circulation is essential for further insight into the dynamics of the vast open ocean regions.

In addition to the surface interactions, vertical mixing processes, and mesoscale motions, the study of the longer-period variations of climate clearly requires consideration of the large-scale dynamics of the complete oceanic circulation. This includes the large-scale pattern of wind-driven and thermohaline currents and their associated horizontal and vertical transports of heat, momentum, and salt. Of particular importance here is the study of the local dynamics of the intense boundary and equatorial currents and the relative roles of inertial and topographic influences. The characteristic variations of these large-scale processes are on time scales of the order of seasons and years in the near-surface waters but may occur in progressively longer time scales in deeper water. The longest oceanic adjustment time associated with the "permanent" ocean circulation is of the order 10^3 years (see Figure 3.3). For climatic variations on these time scales, therefore, the entire water mass of the global ocean must be taken into account.

Modeling the Oceanic Circulation

The systematic examination of the various mechanisms and feedbacks by which the oceanic thermal structure and circulation are maintained on various time scales is largely a task for the future. In this research, it will be necessary to conduct intensive observational programs in order to gain greater understanding of the various oceanic physical processes themselves and to construct numerical models of the oceanic circulation in which these processes are correctly represented.

For climatic studies, it is important that the heat and energy balances of the ocean be modeled correctly over the time and space scales of interest, and this cannot now be said to have been achieved. The classical ocean circulation models, which were initiated in the late 1940's and further developed in the following decades, do account for the gross features of the ocean circulation, such as the major current systems and the large-scale oceanic thermal structure (see Appendix B). But even these features are physically and geometrically distorted by the consideration of only the larger-scale, relatively viscous motions. The commonly used vertical thermal eddy diffusivity in such models is also questionable and may be an order of magnitude too high, as indicated, for example, by recent studies on oceanic tritium concentrations. This alone will produce a distortion of the processes responsible for deep-water formation in such models.

But perhaps more important is the fact that numerical ocean models have not had a sufficiently fine horizontal resolution to portray the mesoscale eddies, either in the open ocean or in the restricted regions of concentrated currents. The accuracy with which the meandering and vortex shedding of boundary currents such as the Gulf Stream or Kuroshio must be modeled, or the resolution required for the transient behavior of the equatorial and Antarctic current systems, depends on the extent to which these features are coupled to the semipermanent or primary current systems themselves and on the time scales under consideration. It is unlikely, however, that these features, or the mesoscale eddies, can be successfully modeled with constant eddy diffusion coefficients.

To study the role of the oceans in climatic change, it is necessary to construct dynamically and energetically correct oceanic general circulation models and to couple them, in appropriate versions, to similarly accurate and compatible atmospheric models. Some experience with simplified coupled models of coarse resolution has already been gained, as discussed in Appendix B. Further tests of coupled models are necessary in which the oceanic mesoscale eddies are resolved, in order that we may understand their role in the oceanic heat balance and their relationship to the climatically important changes of sea-surface temperature. Since computational limitations will likely preclude the resolution of these eddies throughout the world ocean, their successful parameterization will become an important problem for future research.

Of particular importance for climate studies is the construction of an accurate model of the oceanic-surface mixed layer, since all the physical processes in the ocean ultimately exert their influence on the atmosphere through the surface of the sea. Until the dynamics of this oceanic boundary layer are better understood, our ability to model climatic variations on any time scale will remain seriously limited.

SIMULATION AND PREDICTABILITY OF CLIMATIC VARIATION

Climate Modeling Problem

From the above remarks it is clear that the problem of modeling climatic variation is fundamentally one of constructing a hierarchy of coupled atmosphere–ocean models, each suited to the physical processes dominant on a particular time scale. The attack on this problem is now in its infancy. Whether we consider changes of the external boundary conditions or changes of the internally controlled physical processes and feedback mechanisms, we note from Figure 3.3 the wide range of time intervals over which characteristic climatic events occur and that many of these involve interactions among the atmosphere, oceans, ice, and land. Because of the system's nonlinearity, we may expect a broad range of response in both space and time in the individual climatic variables. This is just what the climatic record shows.

To study the relative contribution of individual physical processes to the overall "equilibrium" climatic state, one approach is to test the sensitivity of the statistics generated by a climate model to perturbations in the parameters that influence that particular physical process. In such a modeling program, the effects of changes can first be tested in isolation from other interacting components of the system and then in concert with all known processes in a complete climatic model. In this research, we should not rely exclusively on the general circulation models (GCM's) but should employ a variety of modeling approaches. We note, however, that not only are the GCM's (and the *coupled* GCM's in particular) useful in the calibration of the simpler models, but they are essential to the detailed diagnosis of the shorter-period climatic states that are in approximate statistical equilibrium with slowly changing boundary conditions.

A fundamental approach to the problem of modeling climate and climatic variation must proceed through the consideration of dynamical models of the coupled components of the climatic system. In minimum practical terms, this means the joint atmosphere–ocean system, although for some purposes (such as the behavior of ice sheets and glaciers) the cryosphere must be included as well. Efforts to assemble such models are just getting under way, and their further development is given high priority in the research program recommended in Chapter 6.

Predictability and the Question of Transitivity

It is possible to regard climatic change as a conventional initial/boundary-value problem in fluid dynamics, if we define the climatic system as consisting of the atmosphere, hydrosphere, and cryosphere. In this deterministic view the behavior of the system is governed by the changes of the external boundary conditions (see Figure 3.1). Over relatively short periods, it is even possible to regard the land ice masses as part of the external conditions as well. It is probably *not* possible, however, to remove the hydrosphere from the internal system and still talk meaningfully about climatic *variation*, as the surface layers of the ocean interact with the atmosphere on the shortest time scales associated with climate (see Figure 3.3). Decoupling of the ocean, however, is exactly what has so far been done in conventional atmospheric and oceanic general circulation models, although preliminary efforts to consider the coupled system have been made (see Appendix B).

Even with the atmosphere (together with certain surface effects) regarded as the sole component of the climatic system, and with all external boundary conditions held fixed, there is, in spite of our physical expectations, no assurance that there will be a climate in the sense that time series generated by the atmospheric changes will settle into a statistically steady state; and no assurance that the climate, if it exists, is unique in the sense that the statistics are independent of the initial state. It is therefore useful to define a random time series (or the system generating such a series) as transitive if its statistics (and hence its climatic states) are stable and independent of the initial conditions and as intransitive if not. As shown by Lorenz (1968), nonlinear systems, which are far simpler than the atmosphere, sometimes display a tendency to fluctuate in an irregular manner between two (or more) internal states, while the external boundary conditions remain completely unchanged. This behavior is related to the system's transitivity and is illustrated in Figure 3.4.

Let us assume that two different states of a climatic system are possible at a time $t=0$, such as A and B in Figure 3.4, and let us consider that A is the climatic state that would normally be "expected" under the given constant boundary condition. In a completely transitive system, the climatic state B would approach the state A with the passage of time and eventually become indistinguishable from it. This would correspond to a unique solution for the climate under fixed boundary conditions. In a completely intransitive system, on the other hand, the climatic state B would remain unchanged, and two possible solutions would exist. There would in this case, moreover, be no way in which we could continue to identify the state A as the "normal" or correct solution, as state B would presumably furnish an equally acceptable set of climatic statistics.

A third behavior, however, is perhaps the most interesting of all, and is displayed by an almost-intransitive system. In this case, the system in state B may behave for a while as though it were intransitive, and then at time t_1 shift toward an alternate climatic state A, where it might remain for a further period of time. At time t_2 the system might then return to the original climatic state B, where it could remain or enter into further excursions. The climate exhibited by such a system would thus consist of two (or more) quasi-stable states, together with periods of transition between them. For longer periods of time the system might have stable statistics, but for shorter periods of time it would appear to be intransitive.

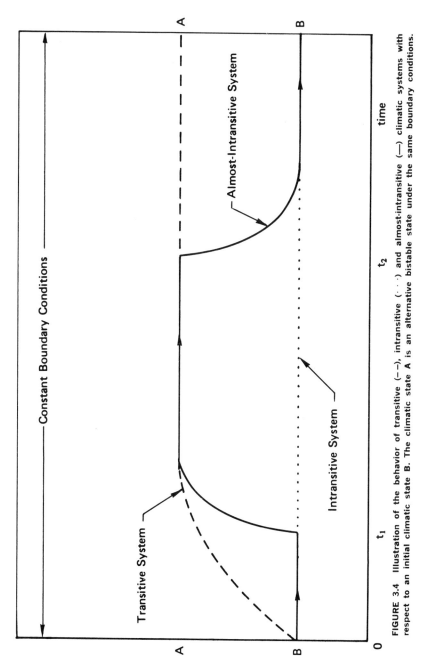

FIGURE 3.4 Illustration of the behavior of transitive (– –), intransitive (· · ·) and almost-intransitive (—) climatic systems with respect to an initial climatic state B. The climatic state A is an alternative bistable state under the same boundary conditions.

Because the atmosphere is constantly subject to disturbances, such as those arising from flow over rough terrain or from the occurrence of baroclinic instability, one might think that it could not be an almost-intransitive system and fail to show greater excursions of annual and decadal climatic states than it does. This depends, of course, on the level of variability associated with individual climatic states and hence on the time interval we select to define the climatic state itself (and on how close neighboring quasi-stable states might be). What may appear to be a climatic transition on one time scale may become the natural noise level of a climatic state defined over a longer interval. This is consistent with many of the climatic records presented in Appendix A.

Even so, it might still be possible for the coupled ocean–atmosphere system or for the coupled ocean–atmosphere–ice system to be almost intransitive. One cannot help but be struck by the appearance of those proxy records that display repeated transitions between two states (see Figures A.13 and A.14 in particular). This evidence suggests that the glacial–interglacial oscillations that have characterized the past million years of the earth's climatic history may be the climatic transitions of an almost-intransitive system. Another possible example of this phenomenon is the irregular and relatively sudden reversal of the earth's magnetic poles. The search for further evidence of this sort in both the paleoclimatic record and in the climatic history generated by numerical models is an important task for future research.

As though the specter of almost-intransitivity were not enough, on the longer-time scales of climatic variation it is equally important to recognize another, potentially serious complication. If it turns out that climatic evolution is influenced to a significant degree by environmental impacts originating *outside* the atmosphere–ocean–cryosphere system, then the predictability of climate will be additionally constrained by the predictability of the environment in a larger sense. This, in turn, could turn out to be the greatest stumbling block of all, as illustrated, for example, by the difficulty of predicting the timing and intensity of volcanic eruptions (which inject radiation-attenuating layers of dust into the upper atmosphere) and by the difficulty of predicting the behavior of the sun itself, which is the ultimate source of the energy driving the climatic system.

As noted earlier, the predictability of climatic variation is constrained by an inherent limitation in the detailed predicability of the atmosphere and ocean. Climatic noise as previously defined thus arises from the unavoidable uncertainty in the determination of the initial state and from the nonlinear nature of the relevant dynamics, as shown, for example, by Lorenz (1969). Fluctuations in the weather for periods beyond a few weeks may therefore be treated in large part as though they were generated by an unpredictable random process. The observed time series of many meteorological variables may be reasonably well modeled by a first-order Markov process with a time (τ)-lagged correlation given by $R(\tau) = \exp(-\nu|\tau|)$, with a constant decay rate ν of the order of 0.3 day^{-1}. The corresponding power spectrum as a function of frequency ω is given by $P(\omega) = A/(\nu^2 + \omega^2)$, where A is a constant. As $\omega \to 0$ for such a spectrum, we have $P(\omega) \to A/\nu^2$, which is a constant, and the very-low-frequency end of the spectrum therefore appears "white." There is thus some contribution to climatic variations on all time scales, no matter how long, arising from the fluctuations of the weather.

While these considerations do not directly address the physical basis of climatic change, they are nevertheless basic to our view of the predictability of climatic change. What parts of climatic variations on various time scales are potentially predictable, and what parts are just climatic noise? In the power spectrum is there potentially predictable variability above the "white," low-frequency end of the daily weather fluctuations, or is it possible that some of the long-term compensation processes, such as those shown in Figure 3.3, might depress the spectrum below its white extension to $\omega = 0$?

The 250-year record of monthly mean temperatures in central England compiled by Manley (1959) shows small lagged correlations significantly above that of weather out to about 6 months, small but perhaps significant lagged correlations at 2 and 4 years, and a generally white spectrum with some evidence of extra variability for periods of a few decades and longer (C. E. Leith, NCAR, Boulder, Colorado, unpublished results). The 6-month lagged correlation may well be a reflection of the role of North Atlantic sea-surface temperature anomalies on the English climate and illustrates the somewhat longer periods of the autovariation of the coupled ocean–atmosphere system over those of the atmosphere alone, as indicated in Figure 3.3. Additional evidence of even longer-period variability is found in the historical and paleoclimatic records (Kutzbach and Bryson, 1974; see Figure A.5). Further studies of this kind should be made with statistical tests not only of the pessimistic null hypothesis that nothing is predictable but also of hypotheses that are framed more optimistically.

Long-Range or Climatic Forecasting

As our understanding of the physical basis of climatic variation grows, we hope to be able to discern the predictable climatic change signal from the unpredictable climatic noise and to describe with some confidence the character of both past and likely future climates. In view of the questions posed by limited predictability, however, this discernment may be limited to those circumstances in which there is a relatively large change in the processes or boundary conditions of the climatic system. The related problem of forecasting specific seasonal and annual climatic variations rests upon the same physical basis and may prove more difficult to solve. To reach these goals will require the coordinated use of all our research tools, whether they be observational, numerical, or theoretical. The capstone of these efforts will be the emergence of an increasingly well-defined and tested theory of climatic variation.

Whether the predictability of climatic change turns out to be lower than many would like to believe or to be limited to a finite range as in the problem of weather forecasting, the quest for understanding must be made. Our recommendations for the research that we believe to be a necessary part of this effort are presented in detail in Chapter 6.

4
PAST CLIMATIC VARIATIONS AND THE PROJECTION OF FUTURE CLIMATES

It is universally accepted that global climate has undergone significant variations on a wide variety of time scales, and we have every reason to expect that such variations will continue in the future. The development of an ability to forecast these future variations, even on time scales as short as one or two decades, is an important and challenging task. Study of the instrumental, historical, and paleoclimatic records not only offers a basis for projection into the future but furnishes insight into the regional effects of global climatic changes. This chapter attempts to summarize our knowledge of past climatic variations and to give some indication of the further research that must be carried out on critical aspects of this subject. Further details of the record of past climates are given in Appendix A.

IMPORTANCE OF STUDIES OF PAST CLIMATES

In order to understand fully the physical basis of climate and climatic variation, we must examine the earth's atmosphere–ocean–ice system under as wide a range of conditions as possible. Most of our notions of how the climatic system works, and the tuning of our empirical and dynamical models, are based on observations of today's climate. In order that these ideas and models may be useful in the projection of future climates, it is necessary that they be calibrated under as wide a range of conditions as possible. The only documented evidence we have of climates under boundary conditions significantly different from today's comes from the paleoclimatic record. It is here that paleoclimatology, in conjunction with climatic modeling, can make an especially valuable contribution to the resolution of the problem of climatic variation.

Modern instrumental data suggest that the atmosphere, at least, may be capable of assuming quite different circulation patterns even with relatively constant boundary conditions and that the resulting variability of climate is strongly dependent on geographical location. Although the data base is much less complete for the oceans, persistent anomalies of sea-surface temperature appear to be related to atmospheric circulation regimes over time scales of months and seasons, and the oceans may show other longer-period variations of which we are now unaware.

In general, the record of past climate indicates that the longer the available record, the more extreme are the apparent climatic variations. An immediate consequence of this "red-noise" characteristic is that the largest climatic changes are *not* revealed by the relatively short record of instrumental observation but must instead be sought through paleoclimatic studies. The record of past climates also contains important information on the range of climatic variability, the mean recurrence interval of rare climatic events, and the tendency for systematic timewise behavior or periodicity. Such climatic characteristics are in general shown poorly, if at all, by the available instrumental records.

RECORD OF INSTRUMENTALLY OBSERVED CLIMATIC CHANGES

Our knowledge of instrumentally recorded climatic variations is largely confined to the record of the past two centuries or so, and it is only in the last 100 years that synoptic coverage has permitted the analysis of the geographical patterns of climatic change over large portions of the globe. It is only during the past 25 years or so that systematic observations of the free atmosphere (mainly in the northern hemisphere) have been made and that regular measurements of the ocean surface waters have been available in even limited regions. Enough data have been gathered, however, to permit the following summary.

A striking feature of the instrumental record is the behavior of temperature worldwide. As shown by Mitchell (1970), the average surface air temperature in the northern hemisphere increased from the 1880's until about 1940 and has been decreasing thereafter (see Figure A.6, Appendix A). Starr and Oort (1973) have reported that, during the period 1958–1963, the hemisphere's (mass-weighted) mean temperature decreased by about 0.6°C. In that period the polar and subtropical arid regions experienced the greatest cooling. The cause of this variation is not known, although clearly this trend cannot continue indefinitely.

It may represent a portion of a longer-period climatic oscillation, although statistical analysis of available records has failed to establish any significant periodic variation between the quasi-biennial cycle and periods of the order of 100 years. The corresponding patterns of precipitation, cloudiness, and snow cover have not been adequately determined, and it would be of great interest to examine the simultaneous variations of oceanic heat storage and imbalances of the planetary radiation budget, once the necessary satellite observations become available.

For the earlier instrumental period, there are scattered records of temperature, rainfall, and ice extent, which clearly show individual years and decades of anomalous character. The only apparent trend is a gradual warming in the European area since the so-called Little Ice Age of the sixteenth to nineteenth centuries.

HISTORICAL AND PALEOCLIMATIC RECORD

Two sources of data are available to extend the record of climate into the pre-instrumental era: historical sources, such as written records, and qualitative observations, which give rise to what may be called "historical" climatic data; and various natural paleoclimatic recorders, which give rise to what may be called "proxy" climatic data.

Nature of the Evidence

The historical record contains much information relating to climate and climatic variation over the past several hundred to several thousand years, and this information should be located, cataloged, and evaluated. Historical data on crop yields, droughts, and winter severity from manuscripts, explorations, and other sources sometimes provide the only available information on the general character of the climate of the historical past. Such information is especially useful in conjunction with selected tree-ring, ice core, and lake sediment data in diagnostic studies of the higher-frequency climatic variability on the time scales of years, decades, and centuries.

For earlier periods, the paleoclimatic record becomes increasingly fragmentary and ultimately nil for the oldest geological periods. But for the past million years, and especially for the past 100,000 years, the paleoclimatic record is relatively continuous and can be made to yield quantitative estimates of the values of a number of significant climatic parameters. Each record, however, must first be calibrated or processed to provide an estimate of the climate. The elevation of an ancient coral reef, for example, is a record of a previous sea level, but before it can be used for paleoclimatic purposes the effect of local crustal movements must be removed. The taxonomic composition of fossil assemblages in marine sediments and the width of tree rings, for example, are known to reflect the joint influence of several ecological factors; here multivariate statistical techniques can be used to obtain estimates of selected paleoclimatic parameters such as temperature and precipitation.

In order to be useful, a proxy data source must also have a stratigraphic character; that is, the ambient values of a climatically sensitive parameter must be preserved within the layers of a slowly accumulating natural deposit or material. Such sources include the sediments left by melting glaciers on land; sediments in peat bogs, lakes, and on the ocean bottom; the layers in soil and polar ice caps; and the annual layers of wood formed in growing trees. Since no proxy source yields as long and continuous a record as would be desired, and the quality of data varies considerably from site to site, a coherent picture of past climate requires the assembly of data from different periods and with different sampling intervals. Such characteristics of the principal proxy data sources are summarized in Table A.1 of Appendix A.

After proxy data have been processed and stratigraphically screened, an absolute chronology must be established in order to date specific features in the climatic record. The most accurate dating technique is that used in tree-ring analysis, where dates accurate to within a single year may be determined over the past several thousand years under favorable conditions. Annually layered lake sediments and the younger ice cores also have a potential dating accuracy to within several years over the past few millenia. For suitable materials, ^{14}C-dating methods extend the absolute time scale to about 40,000 years, with an accuracy of about 5 percent of the material's true age. Beyond the range of ^{14}C dating, the analysis of the daughter products of uranium decay make possible the reconstruction of the climatic chronology of the past million years. For even older records, our chronology is based primarily on potassium–argon radiometric dating as applied to terrestrial lava and ash beds. Stratigraphic levels dated by this method are then correlated with undated sedimentary sequences by the use of paleomagnetic reversals and characteristic floral and faunal boundaries.

Summary of Paleoclimatic History

From the overview of the geological time scale, we live in an unusual epoch: today the polar regions have large ice caps, whereas during most of the earth's history the poles have been ice-free. As shown in Figure A.15 of Appendix A, only two other epochs of extensive continental glaciation have been recorded, one during late Precambrian time (approximately 600 million years ago) and one during Permo-Carboniferous time (approximately 300 million years ago). During the era that followed the Permo-Carboniferous ice age, the earth's climates returned to a generally warmer, nonglacial regime.

Before the end of the Mesozoic era (approximately 65 million years ago) climates were substantially warmer than today. At that time, the configuration of the continents and shallow ocean ridges served to block a circumpolar ocean current in the southern hemisphere. This barrier was formed by South America and Antarctica, which lay in approximately their present latitudinal positions, and by Australia, then a northeastward extension of Antarctica. About 50 million years ago, the Antarctic–Australian passage opened as Australia moved northeastward and as the Indian Ocean widened and deepened. By about 30 million years ago, the Antarctic circumpolar current system was established, an event that may have decisively influenced the subsequent climatic history of the earth.

About 55 million years ago global climate began a long cooling trend known as the Cenozoic climate decline (see Figure A.15). Approximately 35 million years ago there is evidence from the marine record that the waters around the Antarctic continent underwent substantial cooling, and there is further evidence that about 25 million years ago glacial ice occurred along the edge of the Antarctic continent in some locations. During early Miocene time (approximately 20 million years ago) there is evidence that the low and middle latitudes were somewhat warmer. There is widespread evidence of further cooling about 10 million years ago, including the growth of mountain glaciers in the northern hemisphere and substantial growth of the Antarctic ice sheet; this time may be taken as the beginning of the present glacial age. Evidence from marine sediments and from continental glacial features indicates that about 5 million years ago the already substantial ice sheets on Antarctica underwent rapid growth and even temporarily exceeded their present volume. Three million years ago continental ice sheets appeared for the first time in the northern hemisphere, occupying lands adjacent to the North Atlantic Ocean, and during at least the last 1 million years the ice cover on the Arctic Ocean was never significantly less than it is today.

Once the polar ice caps formed, they began a long and complex series of fluctuations in size. Although the earlier record is still not clear, the last million years has witnessed fluctuations in the northern hemisphere ice sheets with a dominant period on the order of 100,000 years (see Figure A.2). These fluctuations may have occurred in parallel with substantial changes in the volume of the West Antarctic ice sheet. By comparison, however, changes in the volume of the ice sheet in East Antarctica were quite small and were probably not synchronous with glaciations in the northern hemisphere.

The major climatic events during the past 150,000 years were the occurrence of two glacial maxima of roughly equal intensity, one about 135,000 years ago and the other between 14,000 and 22,000 years ago. Both were characterized by widespread glaciation and generally colder climates and were abruptly terminated by warm interglacial intervals that lasted on the order of 10,000 years. The penultimate interglacial reached its peak about 124,000 years ago, while the present interglacial (known as the Holocene) evidently had its thermal maximum about 6000 years ago.

Between 22,000 and 14,000 years ago the northern hemisphere ice sheets attained their maximum extent (see Figure A.24). The eastern part of the Laurentide ice sheet (which covered portions of eastern North America) and the Scandinavian ice sheet (which covered parts of northern Europe) both attained their maxima between 22,000 and 18,000 years ago, several thousand years before the maximum of the Cordilleran ice sheet. About 14,000 years ago deglaciation began rather abruptly, and the Cordilleran sheet melted rapidly and was gone by 10,000 years ago. The interval of deglaciation (14,000 to 7000 years ago) was marked in many places by significant secondary fluctuations about every 2000 to 3000 years.

In general, the period about 7000 to 5000 years ago was warmer than today, although the records of mountain glaciers, tree lines, and tree rings reveal that the past 7000 years was punctuated in many parts of the world by colder intervals about every 2500 years, with the most recent occurring about 300 years ago.

For the last 1000 years, the proxy records generally confirm the scattered observations in historical records. The cold period identified above is seen to have consisted of two periods of maximum cold, one in the fifteenth century and another in the late seventeenth century. The entire interval, from about 1430 to 1850, has long been referred to as the Little Ice Age and was characterized in Europe and North America by markedly colder climates than today's.

INFERENCE OF FUTURE CLIMATES FROM PAST BEHAVIOR

Notwithstanding the limitations of our present insight into the physical basis of climate, we are not altogether powerless to make certain inferences about future climate. Beginning with the most conservative approach, we may use the climatic "normal" as a reference for future planning. In this approach, it is tacitly assumed that the future climate will mirror the recently observed past climate in terms of its statistical properties. Depending on the sensitivity of the climate-related application (and on the degree to which the climate is subject to change over a period of years following that for which the "normal" is defined), this kind of inference can be anything from highly useful to downright misleading.

Of the various other approaches to the inference of future climate in which the attempt is made to capture more predictive information than is embodied in the "normal," the most popular have been those based on the supposition that climate varies in cycles. Since the development of modern techniques of time-series analysis, in particular those involving the determination of the variance (or power) spectrum, it has become clear that almost all alleged climatic cycles are either (1) artifacts of statistical sampling, (2) associated with such small fractions of the total variance that they are virtually useless for prediction purposes, or (3) a combination of both. Other approaches, developed to a high degree of sophistication in recent years, include several kinds of nonlinear regression analysis (in which no assumption need be made about the periodic behavior of the climatic time series), which appropriately degenerate to a prediction of the "normal" in cases where the series possess no systematic temporal behavior. The full potential of such approaches is not yet clear but appears promising, at least in certain situations.

Natural Climatic Variations

Regardless of the approach taken to infer future climates, the view that climatic variation is a strictly random process in time can no longer be supported. It has been well established, for example, that many atmospheric variables are serially correlated on time scales of weeks, months, and even years. For the most part such correlations derive from "persistence" and resemble the behavior of a low-order Markov process. Unfortunately, nonrandomness of this kind does not lend itself to long-range statistical prediction. In addition to persistence, long-term trends have a tendency to show up in great number and variety in climatological time series (see Appendix A). Many such trends are now understood to originate from what are called inhomogeneities in the series, as, for example, effects of station relocations, changes in observing procedures, or local microclimatic disturbances irrelevant to large-scale climate. Even after statistical removal of such effects, many "real" trends nonetheless remain and may be recognized as part of a longer-term oscillation of climate. We must, moreover, recognize that the climatic record may also reflect various natural environmental disturbances, such as volcanic eruptions and perhaps changes of the sun's energy output, which are themselves only poorly predictable, if at all. Clearly, a climatic prediction based on the linear extrapolation into the future of a record containing such effects would be highly unrealistic.

The behavior of longer climatic series is seemingly periodic, or quasi-periodic, especially those series that extend into the geological past as reconstructed from various proxy data sources. It is a fundamental problem of paleoclimatology to determine whether this behavior is really what it seems or whether it is an illusion created by the characteristic loss of high-frequency information due to the limited resolving power of most proxy climatic indicators. Illumination of this question would be of great importance to the determination of the basic causes of the glacial–interglacial climatic succession and to the assessment of where the earth stands today in relation to this sequence.

Spectrum analyses of the time series of a wide variety of climatic indices have consistently displayed a "red-noise" character (see, for example, Gilman *et al.*, 1963). That is, the spectra show a gradual increase of variance per unit frequency as one proceeds from high frequencies toward low frequencies. The lack of spectral "gaps" provides empirical confirmation of the lack of any obvious optimal averaging interval for defining climatic statistics. Most spectra of climatic indices are also consistent in displaying some form of quasi-biennial oscillation (see, for example, Brier, 1968; Angell *et al.*, 1969; or Wagner, 1971). This fluctuation is most obvious in the wind data of the tropical stratosphere but also has been shown to be a real if minor feature of the climate at the earth's surface as well.

Time series of some of the longer instrumental records show some suggestion of very-low-frequency fluctuations (periods of about 80 years and longer), but the data sets are not long enough to establish the physical nature and historical continuity of such oscillations. While numerous investigators have reported spectral peaks corresponding to almost all intermediate periods, the lack of consistency between the various studies suggests that no example of quasi-cyclic climatic behavior with wavelengths between those on the order of 100 years and the quasi-biennial oscillation have been unequivocally demonstrated on a global scale. Further discussion of these questions is given in Appendix A (p. 127 *ff*).

Man's Impact on Climate

While the natural variations of climate have been larger than those that may have been induced by human activities during the past century, the rapidity with which human impacts threaten to grow in the future, and increasingly to disturb the natural course of events, is a matter of concern. These impacts include man's changes of the atmospheric composition and his direct interference with factors controlling the all-important heat balance.

Carbon Dioxide and Aerosols

The relative roles of changing carbon dioxide and particle loading as factors in climatic change have been assessed by Mitchell (1973a, 1973b), who noted that these variable atmospheric constituents are not necessarily external parameters of the climatic system but may also be internal variables; for example, the changing capacity of the surface layers of the oceans to absorb CO_2, the variable atmospheric loading of wind-blown dust, and the interaction of CO_2 with the biosphere.

The atmospheric CO_2 concentrations recorded at Mauna Loa, Hawaii (and other locations) show a steady increase in the annual average, amounting to about a 4 percent rise in total CO_2 between 1958 and 1972 (Keeling *et al.*, 1974). The present-day CO_2 excess (relative to the year 1850) is estimated at 13 percent. A comparison with estimates of the fossil CO_2 input to the atmosphere from human activities indicates that between 50 and 75 percent of the latter has stayed in the atmosphere, with the remainder entering the ocean and the biosphere. The CO_2 excess is conservatively projected to increase to 15 percent by 1980, to 22 percent by 1990, and to 32 percent by 2000 A.D. The corresponding changes of mean atmospheric temperature due to CO_2 [as calculated by Manabe (1971) on the assumption of constant relative humidity and fixed cloudiness] are about 0.3°C per 10 percent change of CO_2 and appear capable of accounting for only a fraction of the observed warming of the earth between 1880 and 1940. They could, however, conceivably aggregate to a further warming of about 0.5°C between now and the end of the century.

The total global atmospheric loading by small particles (those less than 5 μm in diameter) is less well monitored than is CO_2 content but is estimated to be at present about 4×10^7 tons, of which perhaps as much as 1×10^7 tons is derived both directly and indirectly from human activities. If the anthropogenic fraction should grow in the future at the not unrealistic rate of 4 percent per year, the total particulate loading of the atmosphere could increase about 60 percent above its present-day level by the end of this century. The present-day anthropogenic particulate loading is estimated to exceed the average stratospheric loading by volcanic dust during the past 120 years but to equal only perhaps one fifth of the stratospheric loading that followed the 1883 eruption of Krakatoa.

The impact of such particle loading on the mean atmospheric temperature cannot be reliably determined from present information. Recent studies indicate that the role of atmospheric aerosols in the heat budget depends critically on the aerosols' absorptivity, as well as on their scattering properties and vertical distribution. The net thermal impact of aerosols on the lower atmosphere (below cloud level) probably depends on the evaporable water content of the surface in addition to the surface albedo. Aerosols may also affect the structure and distribution of clouds and thereby produce effects that are more important than their direct radiative interaction (Hobbs *et al.*, 1974; Mitchell, 1974).

Of the two forms of pollution, the carbon dioxide increase is probably the more influential at the present time in changing temperatures near the earth's surface (Mitchell, 1973a). If both the CO_2 and particulate inputs to the atmosphere grow at equal rates in the future, the widely differing atmospheric residence times of the two pollutants means that the particulate effect will grow in importance relative to that of CO_2.

Thermal Pollution, Clouds, and Surface Changes

There are other possible impacts of human activities that should be considered in projecting future climates. One of these is the thermal pollution resulting from man's increasing use of energy and the inevitable discharge of waste heat into either the atmosphere or the ocean. Although it is not yet significant on the global scale, the projections of Budyko (1969) and others indicate that this heat source may become an appreciable fraction (1 percent or more) of the effective solar radiation absorbed at the earth's surface by the middle of the next century. And if future energy generation is concentrated into large nuclear power parks, the natural heat balance over considerable areas may be upset long before that time. Recent estimates by Haefele (1974) indicate that by early in the next century, the total energy use over the continents will approach 10 percent of the natural heat density of about 50 W/m^2 and that in local industrial areas the man-made energy density may become several hundred times larger.

There is also the possibility that widespread artificial creation of clouds by aircraft exhaust and by other means may induce significant climatic variations, although there is no firm evidence that this has yet occurred. Such effects could serve to increase the already prominent role played by (natural) clouds in the earth's heat balance (see Figure 3.2).

Widespread changes of surface land character resulting from agricultural use and urbanization, and the introduction of man-made sources of evaporable water, may also have significant impacts on future climates. When the surface albedo and surface roughness are changed by the removal of vegetation, for example, the regional climatic anomalies introduced may have large-scale effects, depending on the location and scale of the changes. The creation of large lakes and reservoirs by the diversion of natural watercourses may also have widespread climatic consequences. The list of man's possible future alterations of the earth's surface can be considerably lengthened by the inclusion of more ambitious schemes, such as the removal of ice cover in the polar regions and the diversion of ocean currents. Again, however, it is only through the use of adequately calibrated numerical models that we can hope to acquire the information necessary for a quantitative assessment of the climatic impacts.

5
SCOPE OF PRESENT RESEARCH ON CLIMATIC VARIATION

The overview of the problem of climatic variation presented in the preceding chapters and in the technical appendixes contains only those references to the literature that were helpful in the illustration of a particular viewpoint or necessary to document a specific source of information. In the course of its deliberations, however, the Panel found it necessary to survey present research on climatic variation, as represented by the more recently published literature and by selected ongoing activities. Inasmuch as this information may serve as a useful background to the Panel's recommendations for a national and international program of climatic research, it is summarized here. Even this survey, in which emphasis is given to material published since 1970, must be considered incomplete and necessarily gives precedence to sources of information most readily available to the Panel. Further useful references on various aspects of the problem of climatic variation are to be found in other recent publications (Committee on Polar Research, 1970; National Science Board, 1972; Wilson, 1970, 1971).

CLIMATIC DATA COLLECTION AND ANALYSIS

Here the current status of efforts to assemble climatic data for both the atmosphere and ocean is summarized, and the various observational field programs directed to the collection of specific data of climatic interest are described.

Atmospheric Observations

Climatological data banks are maintained by NOAA's National Climatic Center (NCC) and National Meteorological Center (NMC) and by the military operational weather services, particularly the Air Force's Environmental Technical Applications Center (ETAC) and the Navy's Fleet Numerical Weather Central (FNWC). Using data from these sources, atmospheric data sets specifically for climatic studies have been assembled by the National Center for Atmospheric Research, the Geophysical Fluid Dynamics Laboratory, MIT, and other institutions. Efforts to assemble the rapidly accumulating data from meteorological satellites have also been made by NOAA's National Environmental Satellite Service (NESS) and by the University of Wisconsin. Sustained efforts to assemble and systematically analyze such data for the use of the climatic research community are important tasks for the future.

In addition to the standard compilations of climatological statistics prepared on a routine basis by governmental agencies, new summaries of upper-air data have been prepared (Crutcher and Meserve, 1970; Taljaard et al., 1969); these have permitted the initial construction of the average monthly global distributions of the basic meteorological variables of pressure, temperature, and dew points at selected levels. The analysis of such data in terms of the various statistics of the global circulation is less advanced, although intensive studies of a five-year period in the northern hemisphere have recently been completed (Oort, 1972; Oort and Rasmusson, 1971; Starr and Oort, 1973). These studies provide the most quantitative analyses of the annual climatic variations of the atmosphere yet made, and plans are under way for their extension to additional five-year periods.

Studies of the spatial patterns of observed variability over longer time periods are almost entirely confined to surface variables in the northern hemisphere (Hellerman, 1967; Kutzbach, 1970; Wagner, 1971). Such studies should be extended to other portions of the atmosphere and broadened to include other, less comprehensively observed climatic elements.

An observational analysis of the tropical and equatorial circulation has been completed (Newell et al., 1972), and statistics for the stratospheric climate are becoming increasingly available (Newell, 1972). Comprehensive data on the components of the global atmospheric energy balance are only beginning to be available (Newell et al., 1969), although many rely on older and indirect data for the unobserved elements of the heat balance at the earth's surface (Budyko, 1956, 1963; Lvovitch and Ovtchinnikov, 1964; Möller, 1951; Posey and Clapp,

1964). More recent direct observations from satellites, however, are providing valuable new insight into both the time and space variations of the overall radiation budget of the earth (Vonder Haar and Suomi, 1971) and promise to provide further data of climatic importance as newer and more versatile satellite observational capabilities develop (Chahine, 1974; COSPAR Working Group 6, 1972; Raschke et al., 1973; Smith et al., 1973).

Oceanic and Other Observations

The observational data base for the oceans is much less developed than that for the atmosphere, and oceanic climatic summaries are based largely on observations that are more widely scattered in both space and time. Even for the more traveled parts of the oceans, these data are sufficient only to indicate the average large-scale features of the ocean's structure and circulation (Fuglister, 1960; Hellerman, 1967; Sverdrup et al., 1942; U.S. Navy Hydrographic Office, 1944). Updated compilations of surface stress (Hellerman, 1967)) and sea-surface temperatures (Alexander and Mobley, 1973; Washington and Thiel, 1970) have been made, and summaries of the observed subsurface temperature structure have recently become available for selected oceans (Born et al., 1973; Robinson and Bauer, 1971).

Significant repositories of oceanic data useful for climatic purposes exist at a number of institutions, although a comprehensive oceanic data inventory has not yet been prepared. The Navy's Fleet Numerical Weather Central, the Scripps Institution of Oceanography, the Woods Hole Oceanographic Institution, and the National Marine Fisheries Service, for example, all have specialized oceanographic data banks, as well as data from individual cruises and expeditions. Guides to the oceanic data services of the Environmental Data Service (1973) of NOAA are also available.

An increasing amount of data on oceanic surface conditions is becoming available from satellite observations and other remote-sensing techniques (Munk and Woods, 1973; Shenk and Salomonson, 1972) and offer the promise of routine global monitoring of sea-surface temperature and sea-ice distribution. Satellite data collected by NESS also permit the determination of the snow and ice extent over land; this and other glaciological data are being accumulated by the U.S. Geological Survey. The further extension of oceanographic, sea-ice, and glaciological observations by satellites, buoys, and ships is under active consideration in connection with the FGGE (GARP, 1972; Stommel, 1973) and is part of other large-scale programs as well (International

SCOPE OF PRESENT RESEARCH ON CLIMATIC VARIATION

Decade of Ocean Exploration, 1973; International Glaciological Programme for the Antarctic Peninsula, 1973; Kasser, 1973; Mid-ocean Dynamics Experiment-one, Scientific Council, 1973; Joint U.S. POLEX Panel, 1974).

Observational Field Programs

Many observational data of importance to climatic research have been acquired in special field programs. Some of these are directly related to GARP itself (AMTEX Study Group, 1973; GARP Joint Organizing Committee, 1972, 1973; Houghton, 1974; Kondratyev, 1973), such as the Complete Atmospheric Energetics Experiment (CAENEX), the Air-Mass Transformation Experiment (AMTEX), the GARP Atlantic Tropical Experiment (GATE), and the Arctic Ice Dynamics Joint Experiment (AIDJEX). Others are part of the NSF's International Decade of Ocean Exploration (IDOE) (1973) program (Mid-ocean Dynamics Experiment-one, Scientific Council, 1973), such as the Geochemical Ocean Sections Study (GEOSECS), the Mid-ocean Dynamics Experiment (MODE), the North Pacific Experiment (NORPAX), and the Climate, Long-range Investigation, Mapping, and Prediction (CLIMAP) project. Other field programs are aimed at the monitoring of atmospheric composition and aerosols, such as those of NCAR, the Environmental Protection Agency, and NOAA's Environmental Research Laboratories.

Each of these programs is focused on physical processes of importance in particular geographical regions and is a valuable source of experience and information. There are also international programs of this sort in various stages of planning, such as the Polar Experiment (POLEX) (Joint U.S. POLEX Panel, 1974), the International Glaciological Program for the Antarctic Peninsula (IGPAP) (1973), and the International Southern Ocean Studies (ISOS) programs (ISOS Planning Committee, 1973). Cooperative programs such as these will be necessary for the comprehensive future monitoring, analysis, and modeling of climate and climatic variation.

STUDIES OF CLIMATE FROM HISTORICAL SOURCES

The record of past climates as contained in various historical documents, writings, and archeological material has been increasingly recognized as an important source of information (Bryson and Julian, 1963; Butzer, 1971; Carpenter, 1965; LeRoy Ladurie, 1971; Lamb, 1968, 1972; Ludlam, 1966, 1968). These sources permit the study of historical climates over the past several thousand years. A systematic compilation of material of this sort is being undertaken by the Climatic Research Unit of the University of East Anglia (Lamb, 1973b).

STUDIES OF CLIMATE FROM PROXY SOURCES

The assembly of paleoclimatic information from proxy data sources has attained new importance in recent years with the development of new methods of dating and of new techniques of quantitative climatic inference. In the following, the various efforts in this aspect of climatic research are briefly summarized.

General Syntheses

Two broad surveys of paleoclimatology have appeared in recent years (Funnell and Riedel, 1971; Schwartzbach, 1961), along with textbooks (Flint, 1971; Washburn, 1973) and symposia (Black et al., 1973; Turekian, 1971), which emphasize the glacial processes during the late Cenozoic period. Other recent paleoclimatological syntheses have been concerned with the broad range of Quaternary studies (Wright and Frey, 1965), with the relationships between Pleistocene geology and biology (Butzer, 1971; West, 1968), and with more recent paleoclimatic fluctuations from a meteorological viewpoint (Lamb, 1969). The review of the full range of paleoclimatic events on all time scales given in Appendix A of this report has been made possible by the recent application of improved dating methods to the stratigraphic record of ocean sediments and uplifted reefs. This synthesis illustrates the essential need for an accurate time scale in the interpretation of proxy climatic data.

Chronology

The methods of dendrochronology (Ferguson, 1970; LaMarche and Harlan, 1973), the radiocarbon method (Olsson, 1970; Wendland and Bryson, 1974), and other isotopic dating methods have recently been used to infer the chronology of climate over the past several hundred thousand years (Broecker and van Donk, 1970; Matthews, 1973; Mesolella et al., 1969). Biostratigraphic and paleomagnetic correlations between the marine and continental records have provided a reasonably accurate chronology of the past 60 million years by the use of potassium-argon and other isotopes (Berggren, 1971, 1972; Hays et al., 1969; Kukla, 1970; Ruddiman, 1971; Sancetta et al., 1973; Shackleton and Kennett, 1974a, 1974b; Shackleton and Opdyke, 1973).

Monitoring Techniques

Following the initial efforts to estimate paleotemperatures from isotopic time series (Emiliani, 1955, 1968), recent work has made it possible to separate the effects of temperature from those of ice-volume change (Shackleton and Opdyke, 1973). Multivariate statistical techniques have recently been developed that permit the quantitative estimation of climatic parameters from the concentration of fossil plankton in deep-sea sediments (Imbrie and Kipp, 1971; Imbrie *et al.*, 1973; Kipp, 1974), the growth record of tree rings (Fritts *et al.*, 1971), and the continental distribution of fossil pollen (Webb and Bryson, 1972). These methods have since been applied to the reconstruction of paleo-ocean temperatures (Luz, 1973; McIntyre *et al.*, 1972a; Pisias *et al.*, 1973; Sachs, 1973), as well as to pressure and precipitation anomalies (Fritts, 1972). Isotopic studies of cores taken in the polar ice caps provide measures of the air temperature at the time of ice formation (Dansgaard *et al.*, 1971). Further refinements of such monitoring techniques will help to fill in the paleoclimatic record, especially when several independent methods are available for the same period.

Proxy Data Records and Their Climatic Inferences

Proxy data come from a wide variety of sources; potentially, any biological, chemical, or physical characteristic that responds to climate may provide proxy data useful in the reconstruction of past climates. One of the more prolific sources of long-term climatic information has been the extensive collection of deep-sea cores, obtained routinely over the years on various oceanographic expeditions and more recently from the Deep-Sea Drilling Project (Douglas and Savin, 1973; Shackleton and Kennett: 1974a, 1974b). Analysis of the fossil flora and fauna in such cores, with chronology provided from their isotopic content and paleomagnetic stratigraphy, has been performed for all the principal oceans of the world (Emiliani, 1968; Gardner and Hays, 1974; Hunkins *et al.*, 1971; Imbrie, *et al.*, 1973; Kellogg, 1974; Kennett and Huddlestun, 1972; Moore, 1973) and provides a preliminary documentation of the temperature and large-scale displacements of the surface waters during the last few hundred thousand years (McIntyre *et al.*, 1972b; Shackleton and Opdyke, 1973). Other characteristics of the sediment cores, such as the presence of volcanic ash (Ruddiman and Glover, 1972), also indicate climatically important events, as well as providing valuable core dating horizons. For periods of particular interest, such as the glacial maximum of about 18,000 years ago, detailed reconstructions of seasonal sea-surface temperature and salinity have been made for the North Atlantic (McIntyre et al., 1974) and more recently have been extended to the world ocean under the CLIMAP program.

The concentration of fossil pollen and the record of soil types in relatively undisturbed continental sites is another source of proxy data on terrestrial paleoclimates. In recent years, pollen data have been analyzed from a number of continental areas (Bernabo et al., 1974; Davis, 1969; Heusser, 1966; Heusser and Florer, 1974; Livingstone, 1971; Swain, 1973; Tsukada, 1968; van der Hammen et al., 1971) and provide a preliminary documentation of the surface vegetational changes during the late Cenozoic and Quaternary periods (Leopold, 1969; Wright, 1971). Soil records have been studied less extensively but provide corroborative evidence of surface climatic conditions (Frye and Willman, 1973; Kukla, 1970; Sorenson and Knox, 1973).

In many ways analogous to the records from deep-sea cores, proxy climatic data from ice cores have recently been obtained from sites in both Antarctica and Greenland (Dansgaard et al., 1969, 1971). Such ice-core records provide a detailed history of atmospheric conditions over the ice during the last hundred thousand years (Dansgaard et al., 1973; Johnsen et al., 1972; Langway, 1970). The drilling of deeper cores are planned, and their analysis and correlation with other proxy data will contribute significantly to the reconstruction of global climatic history.

Further climatic inferences are obtained from proxy data on marine shorelines. By assembling data on dated terraces at selected continental and island sites, and with the necessary adjustments for eustatic changes in the earth's crust, the record of sea-level variations over the last 150,000 years is becoming established (Bloom, 1971; Currey, 1965; Matthews, 1973; Mesolella et al., 1969; Milliman and Emery, 1968; Steinen et al., 1973; Walcott, 1972), particularly as regards the timing of high stands.

Closely related to the questions of ice, soil, and sea-level changes are the proxy data from glacial fluctuations themselves. Considerable attention has been given in recent years to the reconstruction of the glacial history of the most recent major ice age in North America (Andrews et al., 1972; Black et al., 1973; Dreimanis and Karrow, 1972; Frye and Willman, 1973; Paterson, 1972; Porter, 1971; Richmond, 1972), as well as the relatively small but significant fluctuations in mountain glaciers over the past 10,000 years (Denton and Karlén, 1973). Although local glacial margins fluctuate primarily in response to the glacier's net mass accumulation, their overall pattern provides evidence of larger-scale and longer-period climatic responses. When these changes are combined with the more limited data on the glacial history of the Antarctic ice sheet, a number of worldwide relationships in the major fluctuations of glacial extent begin to emerge (Denton et al., 1971; Hughes, 1973).

In the postglacial period, important proxy data on climatic variations over the continents also come from the records of tree rings and closed-basin lakes. Both of these features respond directly to the hydrologic and thermal balances at the surface and when properly calibrated for local effects can provide a record of climate over thousands of years. With the introduction of new dating and analysis methods, the records of tree-ring width variations from both living and fossil trees provide an annually integrated record of climatic changes, especially in arid regions (Ferguson, 1970; Fritts, 1971, 1972; LaMarche, 1974; LaMarche and Harlan, 1973). The radiocarbon dating of debris in selected arid lakes provides further evidence of climatic variations, particularly as they affect the local water balance (Broecker and Kaufman, 1965; Butzer et al., 1972; Farrand, 1971).

Institutional Programs

Much of the present research on paleoclimates is performed in conjunction with other glaciological and geological programs, such as those of the U.S. Geological Survey, the Illinois Geological Survey, the Lamont-Doherty Geological Observatory of Columbia University, and the Army's Cold Regions Research and Engineering Laboratory. Other efforts are conducted within the larger oceanographic research laboratories, such as the Scripps Institution of Oceanography of the University of California, the Woods Hole Oceanographic Institution, the U.S. Naval Oceanographic Laboratory, and the marine research laboratories of the University of Miami, the University of Rhode Island, and Oregon State University. In recent years, more specialized paleoclimatic research efforts have been developed at a number of other universities, joining the long-established Laboratory of Tree-ring Research of the University of Arizona and the Institute for Polar Research at The Ohio State University. These include the Quaternary Research Centers at the University of Washington and the University of Maine, the Center for Climatic Research at the University of Wisconsin, the Institute of Arctic and Alpine Research at the University of Colorado, and the paleoclimatic research programs in the geology and geophysics departments of Brown University and Yale University.

Notable among the many cooperative activities of these and other institutions are the NSF's IDOE programs, including the CLIMAP and NORPAX projects. Such cooperative programs have been instrumental in developing an effective collaboration among the paleoclimatic, oceanographic, and meteorological research communities and should be broadened in the future.

PHYSICAL MECHANISMS OF CLIMATIC CHANGE

Although the problem of climatic change has been the subject of speculation for over a century, recent research has concentrated on the study of specific physical processes and on the interactions among various components of the climatic physical system. Here the more recent of such efforts are briefly surveyed, together with a review of associated empirical, diagnostic, and theoretical studies.

Physical Theories and Feedback Mechanisms

Of particular interest in the problem of climatic change is the question of the cause of the ice ages. Among the recent attempts to answer this question are hypotheses that focus upon the roles of sea ice (Donn and Ewing, 1968) and ice shelves (Wilson, 1964), the carbon dioxide balance (Plass, 1956), and the ocean's salinity (Weye, 1968). Other hypotheses emphasize the roles of variations of external boundary conditions, particularly the incoming solar radiation (Alexander, 1974; Budyko, 1969; Clapp, 1970; Manabe and Wetherald, 1967) and the volcanic dust loading of the atmosphere (Lamb, 1970).

It is generally believed that the astronomical variations of seasonal solar radiation play a role in longer-period climatic changes (Milankovitch, 1930; Mitchell, 1971b; Vernekar, 1972), although there is no agreement on the physical mechanisms involved. Recent studies have also been made of the long-standing question of possible short-term relationships between the climate and solar activity itself (Roberts, 1973; Roberts and Olson, 1973). Other hypotheses of climatic change reckon with the possibility that much of the observed variations of climate are essentially the result of the natural, self-excited variability of the internal climatic system (Bryson, 1974; Mitchell, 1966, 1971b; Sawyer, 1966).

Of the many feedback processes involved in climate (Schneider and Dickinson, 1974) the role of aerosols has recently received particular attention (Chylék and Coakley, 1974; Joseph et al., 1973; Mitchell, 1971a, 1974; Rasool and Schneider, 1971; Schneider, 1971). Although our knowledge of the physical properties and global distribution of aerosols is limited, these studies indicate that the climatic effects may be substantial (Rasool and Schneider, 1971; Yamamoto and Tanaka, 1972). Several research programs on aerosols are under way, including the Global Atmospheric Aerosol Research Study (GAARS) of NCAR and the Soviet CAENEX program (Kondratyev, 1973) previously noted. Attention has also been focused on the regulatory roles of cloudiness (Cox, 1971; Mitchell, 1974; Schneider, 1972) and air–sea interaction (Namias, 1973; White and Barnett, 1972) in the global climatic system. In both cases, however, an adequate quantitative understanding has not yet been achieved.

Diagnostic and Empirical Studies

Related to the search for physical climatic theories and mechanisms are many empirical and diagnostic studies of various aspects of climatic change. Particular attention has been given to the analysis of the large-scale variations of the atmospheric circulation that have been observed during the past few decades (Angell *et al.,* 1969; Bjerknes, 1969; Davis, 1972; Namias, 1970; Wahl, 1972; Wahl and Lawson, 1970; White and Walker, 1973) and to their relationship to regional anomalies of temperature and rainfall (Landsberg, 1973; Namias, 1972b; Winstanley, 1973a, 1973b). Satellite observations of the large-scale variations of surface albedo and seasonal snow cover have brought new attention to these features of the climatic system (Kukla and Kukla, 1974; Wagner, 1973), as well as necessitating a significant revision of the atmospheric radiative energy budget (London and Sasamori, 1971) and the estimated oceanic energy transport (Vonder Haar and Oort, 1973).

Several recent diagnostic and empirical studies have also focused on aspects of the atmosphere–ocean interaction on seasonal, annual, and decadal time scales (Lamb and Ratcliffe, 1972; Namias, 1969, 1971b, 1972a) and have prompted new attention to their relevance to long-range forecasting (Ratcliffe, 1973; Ratcliffe and Murray, 1970). The larger-scale variations of ocean-surface temperature and sea level have also been studied and have led to the identification of apparent teleconnections with the atmospheric circulation (Namias, 1971a; Wyrtki, 1973, 1974).

New studies of mesoscale oceanic features have been made (Bernstein, 1974) and provide further evidence of the dominance of this scale in the oceanic energy spectrum (in agreement with the preliminary results of the MODE program). Other oceanic studies have concentrated on the empirical evaluation of the turbulent fluxes of momentum, heat, and water vapor across the air–sea interface (Holland, 1972; Paulson et al., 1971, 1972). The difficulties of estimating the transport of even the strongest ocean currents or the heat balance over ice-covered seas with the present data base have also received renewed emphasis (Fletcher, 1972; Niiler and Richardson, 1973; Reid and Nowlin, 1971).

Predictability and Related Theoretical Studies

An important problem in climatic variation is the determination of the degree of predictability that is inherent in the natural system, as well as that which is achievable by simulation. A number of recent studies of simplified models have shown that multiple climatic solutions may exist under the same external conditions (Budyko, 1972b; Faegre, 1972; Lorenz, 1968, 1970) in a manner suggestive of certain features of the observed climatic record. There is also evidence from simplified models that the completely accurate specification of a climatic state is not achievable in any case, because of the same kind of nonlinear error growth that limits the accuracy of weather prediction (Fleming, 1972; Houghton, 1972; Leith, 1971; Lorenz, 1969; Robinson, 1971a).

Analyses of selected climatic time series indicate only limited predictability on yearly and perhaps decadal time scales (Kutzbach and Bryson, 1974; Leith, 1973; Lorenz, 1973; Vulis and Monin, 1971), while the general white-noise character of higher-frequency fluctuation has been confirmed in model simulations (Chervin et al., 1974). Further studies of climatic predictability are needed in order to identify both the intrinsic and practical limits of climatic prediction.

NUMERICAL MODELING OF CLIMATE AND CLIMATIC VARIATION

The accurate portrayal of global climate is the scientific goal of much of the atmospheric and oceanic numerical modeling effort now under way (Smagorinsky, 1974). When such models are coupled, the direct numerical simulation of at least the shorter-period climatic variations becomes a realistic possibility. The study of longer-period climatic variations, however, may require the construction of increasingly parameterized models. Here the more recent modeling research in both of these approaches is briefly reviewed.

Atmospheric General Circulation Models and Related Studies

Studies with global atmospheric general circulation models (GCM's) have focused on the simulation of seasonal climate, with emphasis on

SCOPE OF PRESENT RESEARCH ON CLIMATIC VARIATION 57

the analysis of the surface heat and hydrologic balances (Gates, 1972; Holloway and Manabe, 1971; Kasahara and Washington, 1971; Manabe, 1969a, 1969b; Manabe et al., 1972; Somerville et al., 1974). As described more fully in Appendix B, simulations of average January climate have now been achieved by several GCM's. Although additional global GCM's are under development (Corby et al., 1972), only two at this writing have been integrated over time longer than one year (Manabe et al., 1972, 1974b; Mintz et al., 1972).

Global atmospheric models have also recently been applied to the simulation of specific regional circulations, such as those in the tropics (Manabe et al., 1974). In such applications the model's parameterization of processes in the surface boundary layer is of particular importance (Deardorff, 1972; Delsol et al., 1971; Sasamori, 1970). Considerable recent interest has also been shown in the simulation of stratospheric climate with global GCM's (Kasahara and Sasamori, 1974; Kasahara et al., 1973; Mahlman and Manabe, 1972). An overview of global atmospheric (and oceanic) GCM's is given in Appendix B; more detailed reviews of these and other models have recently been prepared (Robinson, 1971b; Schneider and Kellogg, 1973), while others are in preparation (GARP Joint Organizing Committee, 1974; Schneider and Dickinson, 1974).

Statistical–Dynamical Models and Parameterization Studies

Research on the development of dynamical climate models (in which the transfer properties of the large-scale eddies are statistically parameterized rather than resolved as in the GCM's) has accelerated in recent years (Willson, 1973). These models include those that address only the surface heat balance (Budyko, 1969; Faegre, 1972; Sellers, 1969, 1973), those that consider the time-dependent zonally averaged motion (MacCracken, 1972; MacCracken and Luther, 1973; Saltzman and Vernekar, 1971, 1972; Wün-Nielsen, 1972; Williams and Davies, 1965), and those in which the statistical eddy fluxes are represented in terms of the large-scale motions themselves (Dwyer and Petersen, 1973; Kurihara, 1970, 1973). A key problem in such models is the correct parameterization of the heat and momentum transports by the large-scale eddies. While a completely adequate formulation has not yet been achieved, research is continuing by a variety of methods (Clapp, 1970; Gavrilin and Monin, 1970; Green, 1970; Saltzman, 1973; Smith, 1973; Stone, 1973). Because of the generally longer time scales involved in the oceanic general circulation, relatively less attention has been given to the corresponding formulation of statistical–dynamical ocean models (Adem, 1970; Petukhov and Feygel'son, 1973; Pike, 1972). This problem, however, will assume greater importance with the increased development of coupled ocean–atmosphere systems reviewed below.

Oceanic General Circulation Models

Although generally less advanced than their atmospheric counterparts oceanic GCM's have recently been developed to the point where successful simulations of the seasonal variations of ocean temperature and currents have been achieved in both idealized basins (Bryan, 1969; Bryan and Cox, 1968; Haney, 1974) and in selected ocean basins with realistic lateral geometry (Cox, 1970; Galt, 1973; Holland and Hirschman, 1972; Huang, 1973). The numerical simulation of the complete world ocean circulation has only recently been achieved with baroclinic models (Alexander, 1974; Cox, 1974; Takano et al., 1973); this shows significant improvement over earlier global simulations with homogeneous wind-driven models (Bryan and Cox, 1972; Crowley, 1968). As noted earlier, such models have not yet been able to resolve the energetic oceanic mesoscale eddies, although a number of experimental calculations to that end are under way.

Recent studies have also shown the importance of improving the models' treatment of the oceanic surface mixed layer (Bathen, 1972; Denman, 1973; Denman and Miyake, 1973) and sea ice (Maykut and Untersteiner, 1971) and of incorporating bottom topography (Holland, 1973; Rooth, 1972) and the abyssal water circulation (Kuo and Veronis, 1973).

Coupled General Circulation Models

Although preliminary numerical calculations with a model of the coupled atmosphere–ocean circulation were performed several years ago (Manabe and Bryan, 1969; Wetherald and Manabe, 1972), it is only recently that a truly globally coupled model has been achieved (Bryan et al., 1974; Manabe et al., 1974a). These calculations underscore the importance of the ocean's participation in the processes of air–sea interaction and in the maintenance of large-scale climate. These and other such coupled models now under construction will lay the basis for the systematic exploration of the dynamics of the atmosphere–ocean system and its role in climatic variation. The necessary calibration and testing of coupled GCM's will require a broad data base and access to the fastest computers available.

APPLICATIONS OF CLIMATE MODELS

The uses of climate models extend across a wide range of applications, including the reconstruction of past climates and the projection of future climates. Here the more recent use of models for such studies is briefly reviewed, as distinguished from the research on model design and calibration reviewed above.

Simulation of Past Climates

By assembling the boundary conditions appropriate to selected periods in the past, numerical models may be applied to the simulation of paleoclimates. The climate of the last ice age has recently received increased attention, both through the application of parameterized and empirical models (Alyea, 1972; Lamb and Woodroffe, 1970; MacCracken, 1968) and through the use of atmospheric GCM's (Kraus, 1973; Williams et al., 1973). In the latter case, the specification of the distribution of glacial ice and sea-surface temperature represents a strong thermal control over the simulated climate. In order to provide realistic information on the near-equilibrium ice-age climatic state, these conditions should be constructed on the basis of the appropriate proxy climatic records, while other portions of this same paleoclimatic data base serve as verification. An initial effort of this sort is now under way as part of the CLIMAP program.

At the present time, the simulation of the time-dependent evolution of past climates over thousands of years can only be achieved with the more highly parameterized models; the design and calibration of such models of the air–sea–ice system are largely tasks for the future.

Climate Change Experiments and Sensitivity Studies

Numerical climate models also permit the examination of the climatic consequences of a wide variety of possible changes in the physical system and its boundary conditions; such studies, in fact, are a primary motivation for the development of the climatic models themselves. As previously noted, a number of experiments on the effect of solar radiation changes have been performed with simplified models (Budyko, 1969; Manabe and Wetherald, 1967; Schneider and Gal-Chen, 1973; Sellers, 1969, 1973), and further studies of this kind with global models are under way. A number of recent experiments have been made with atmospheric GCM's on the effects of prescribed sea-surface temperature anomalies on the large-scale atmospheric circulation (Houghton et al.,

1973; Rowntree, 1972; Spar, 1973a, 1973b), while others have been concerned with the climatic effects of thermal pollution (Washington, 1972) and of sea ice (Fletcher 1972).

Although these experiments indicate that the models display a response over several months' time to small changes in the components of the surface heat balance, their longer-term climatic response is not known. Such experiments serve to emphasize the need for extended model integrations, preferably with coupled models, and underscore the importance of determining the models' sensitivity and the consequent noise levels in model-generated climatic statistics. The reduction of this climatic noise has an important bearing on the determination of the significance of climatic variations (Chervin et al., 1974; Gates, 1974; Gilman et al., 1963; Leith, 1973). This question is also closely related to the problem of long-range or climatic prediction (Brier, 1968; Kukla et al., 1972; Lamb, 1973a).

Studies of the Mutual Impacts of Climate and Man

Although the influence of man's activities on the local climate has long been recognized, renewed attention has been given in recent years to the possibility that man's increasingly extensive alteration of the environment may have an impact on the large-scale climate as well (Sawyer, 1971). Here the more recent of such studies are briefly reviewed, along with studies of the parallel problem of climate's impact on man's activities themselves.

Aside from the numerical simulations of anthropogenic climatic effects noted earlier, there have been a number of recent studies of the climatic consequences of atmospheric pollution (Bryson and Wendland, 1970; Mitchell, 1970, 1973a, 1973b; Newell, 1971; Yamamoto and Tanaka, 1972) and of the possible effects of man's interference with the surface heat balance, primarily through changes of the surface land character (Atwater, 1972; Budyko, 1972a; Landsberg, 1970; Sawyer, 1971). Aside from local climatic effects, such as those due to urbanization, these studies have not yet established the existence of a large-scale anthropogenic climatic impact (Machata, 1973). Like their numerical simulation counterparts, such studies are made more difficult by the high levels of natural climatic variability and by the lack of adequate observational data.

A longer-range question receiving increased attention is the problem of disposing of the waste heat that accompanies man's production and consumption of energy. When projected into the next century, this effect poses potentially serious climatic consequences and may prove

SCOPE OF PRESENT RESEARCH ON CLIMATIC VARIATION 61

to be a limiting factor in the determination of acceptable levels of energy use (Haefele, 1973; Lovins, 1974). These and other aspects of man's impact on the climate have been considered extensively in the SCEP and SMIC reports (Wilson, 1970, 1971).

Recent attention has also focused on the effects of climatic variations on man's economic and social welfare. From a general awareness of these effects (Budyko, 1971; Johnson and Smith, 1965; Maunder, 1970) research has turned to the representation of climatic anomalies in terms of the associated agricultural and commercial impacts (Pittock, 1972) and to the development of climatic impact indices (Baier, 1973). Further studies are necessary in order to develop comprehensive climatic impact simulation models, with both diagnostic and predictive capability.

6
A NATIONAL CLIMATIC RESEARCH PROGRAM

While there is ample evidence that past climatic changes have had profound effects on man's activities, future changes of climate promise to have even greater impacts. The present level of use of land for agriculture, the use of water supplies for irrigation and drinking, and the use of both airsheds and watersheds for waste disposal is approaching the limit. A change of climate, even if sustained only for a few years' time, could seriously disrupt this use pattern and have far-reaching consequences to the national economy and well-being. To this vulnerability to natural climatic changes we must add the increasing possibility that man's own activities may have significant climatic repercussions.

If we are to react rationally to the inevitable climatic changes of the future, and if we are ever to predict their future course, whether they are natural or man-induced, a far greater understanding of these changes is required than we now possess. It is, moreover, important that this knowledge be acquired as soon as possible. Although much has been accomplished, and further research is under way on many problems (as summarized in Chapter 5), the mechanics of the climatic system is so complex, and our observations of its behavior so incomplete, that at present we do not know what causes any particular climatic change to occur.

Our response to this state of affairs is the recommendation of an integrated research program to contain the observational, analytical, and research components necessary to acquire this understanding.

Heretofore the many pieces of the climatic puzzle have been considered in relative isolation from each other, a subdivision that is natural to the traditional scientific method. We believe, however, that the time has now come to initiate a broad and coordinated attack on the problem of climate and climatic change. Such a program should not stifle the development of new and independent lines of attack nor seek to assemble all efforts under a single authority. On the contrary, its purpose should be to provide a coordinating framework for the necessary research on all aspects of this important problem, including the strengthening of those efforts already under way as well as the initiation of new efforts. Only in this manner can our limited resources be used to maximum benefit and a balanced and coherent approach maintained.

THE APPROACH

From the summary of recent and current research on climate and climatic variation presented in Chapter 5, it is clear that considerable effort has been devoted to this problem. It is also clear that much remains to be done. As an approach to the research program itself, we here attempt to summarize what is now known and to identify those elements that now make a greatly expanded effort both feasible and desirable.

What Climatic Events and Processes Can We Now Identify?

From the analysis of accumulated instrumental climatic data, we can identify some of the major characteristics of the climatic changes of the past few decades. These include the presence of seasonal and annual circulation anomalies over large regions of the earth, together with some longer-term trends. More recent satellite observations have documented changes in worldwide cloudiness, snow cover, and the global radiation balance and have served to emphasize the climatic role of the oceans. Although the necessary oceanic measurements have not yet been made, satellite observations (together with atmospheric data) indicate that the oceans accomplish between one third and one half of the total annual meridional heat transport.

From the analysis of selected paleoclimatic data, it appears that ancient climates have been somewhat similar in behavior to the present-day climate, although the resolution is poorer. These data also suggest the presence of seemingly quasi-periodic climatic fluctuations on time scales of order 100,000 years, associated with the earth's major glaciations.

From the solutions of numerical general circulation models, we can identify a number of important physical elements in the maintenance of global climate. Primary among these is the role played by convective motions in the vertical heat flux and by the transfers of heat at the ocean surface. Climate models also show that the climate is sensitive to the extent of cloudiness and to the surface albedo. Recent solutions of global atmospheric models have shown that the accuracy of the simulations of cloudiness and precipitation is more difficult to establish than the average seasonal distribution of the large-scale patterns of pressure, temperature, and wind, which are simulated with reasonable accuracy (see Appendix B). This may be due to the prescription of the sea-surface temperature in the atmospheric models, serving to mask errors in the models' heat balance.

Less experience has been gained with oceanic general circulation models, although they are capable of portraying the large-scale thermal structure of the oceans and the distribution of the major current systems when subject to realistic (atmospheric) surface boundary conditions. These and other models are just beginning to identify the energetic mesoscale eddy, which in some ways appears to be the oceanic counterpart of the transient cyclones and anticyclones in the atmosphere.

From the analysis of a variety of climate models, as well as from the analysis of climatic data, we can identify a number of links or processes in the phenomenon of climatic change. On at least the shorter climatic time scales, the climatic system is regulated by a number of feedback mechanisms, especially those involving cloudiness, surface temperature, and surface albedo. Underlying these effects is the increasing evidence that large-scale thermal interactions between the ocean and atmosphere are the primary factor in climatic variations on time scales from months to millenia. These interactions must be examined in coupled ocean–atmosphere models, whose development has just begun. The role of the oceans in the climatic system raises the possibility of some degree of useful predictability on, say, seasonal or annual time scales and is an obviously important matter for further research.

From the analysis of the limited data available, we can identify a number of areas in which man's actions may be capable of altering the course of climatic change. Chief among these is interference with the atmospheric heat balance by increasing the aerosol and particulate loading and increasing the CO_2 content of the atmosphere by industrial and commercial activity. While present evidence indicates that these are not now dominant factors, they may become so in the future. To these we must also add the possibility of man's direct thermal interference with climate by the disposal of large amounts of waste heat into the atmosphere and ocean. Although important large-scale thermal pollution effects of this sort do not appear likely before the middle of the next century, they may eventually be the factor that limits the climatically acceptable level of energy production.

Why Is a Program Necessary?

Although the conclusions identified above represent important research achievements, they are nevertheless concerned with separate pieces of the problem. What we *cannot* identify at the present time is how the complete climatic system operates, which are its most critical and sensitive parts, which processes are responsible for its changes, and what are the most likely future climates. In short, while we know something about climate itself, we know very little about climatic *change*.

From among the present activities we can identify important problems requiring further research. In general, these concern new observations and the further analysis of older ones, the design of improved climatic models of the atmosphere and ocean, and the simulation of climatic variations under a variety of conditions for the past, present, and future. As we attempt this research on a global scale, it becomes increasingly important that we ensure the smooth flow of data and ideas, as well as of resources, among all parts of the problem. The attention devoted in each country (and internationally through GARP) to the improvement of weather forecasting (a problem whose physical basis is reasonably well understood) must be matched by a program devoted to climate and climatic variation, a problem whose global aspects are even more prominent and whose physical basis is not at all well understood. The need for a broad, sustained, and coordinated attack is therefore a fundamental reason for a climatic research program.

Other circumstances also indicate that a major research program on climatic change is both timely and necessary. First, for the past few years we have had available to us the unprecedented observational capability of meteorological satellites. This capability has steadily increased from the initial observations of the cloudiness, radiation budget, and albedo to include the vertical distribution of temperature and moisture, the extent of snow and ice, the sea-surface temperature, the presence of particulates, and the character of the land surface. The regular global coverage provided by such satellites clearly constitutes an observational breakthrough of great importance for climatic studies.

Second, the steady increase in the speed and capacity of computers, which has been taking place since their introduction in the 1950's, has reached the point where numerical integration of global circulation models over many months or even years is now practical. Such calculations, along with the associated data processing, will form the quantitative backbone of climatic research for many years to come, and their feasibility clearly constitutes a computational breakthrough. This computing capability, as represented, for example, by machines of the TI-ASC or ILLIAC-4 class, will permit extensive experimentation for the first time with the coupled global climatic system.

Finally, the recent development of unified physical models of the coupled ocean–atmosphere may itself be viewed as a modeling breakthrough of great importance. Up to now either the atmosphere or the ocean has been considered as a separate entity in global modeling, and their solutions have consequently described a sort of quasi-equilibrium climate. The simulation of climatic variation with these models, on the other hand, is just now beginning. A future modeling breakthrough of equally great importance will be the successful parameterization of the eddy transports of baroclinic disturbances in the atmosphere and in the ocean.

Aside from the practical importance (or even urgency) of the climatic problem, the breakthroughs noted above indicate that a time is at hand during which progress will be in proportion to our efforts. By coordinating these efforts into a coherent research program, we may therefore expect to achieve significantly greater understanding of climatic variation.

THE RESEARCH PROGRAM (NCRP)

We have here assembled our specific recommendations for the *data*, the *research*, and the *applications* that we believe constitute the needed elements of a comprehensive national research program on climatic change. We recognize that some of the elements of this program require considerable further development and coordination. We also recognize that some of the recommended efforts are already under way or are planned by various groups, but we believe that their identification as parts of a coherent program is both valuable and necessary. Our recommendations for the planning and execution of this program are given later in this chapter, including those items on which we urge immediate action.

Data Needed for Climatic Research

The availability of suitable climatic data is essential to the success of climatic analysis and research, and such data are an integral part of the overall program. The needed data are discussed below in terms of a subprogram for climatic data assembly and analysis and a subprogram for climatic index monitoring. These are the efforts that we believe to be necessary to make the store of climatic data more useful to the climatic research community and to ensure the systematic collection of the needed climatic data in the future.

Climatic Data Analysis

Instrumental Data Instrumental observations of the atmosphere adequate to depict even a decadal climatic variation are available only for about the last half century for selected regions of the northern hemisphere, and the observational coverage of the oceans is even poorer in both space and time (see Appendix A). In order to assess more accurately the present data base of conventional observations and the needed extensions of such data, a number of efforts should be undertaken:

A worldwide *inventory of climatic data* should be taken to determine the amount, nature, and location of past and present instrumental observations of the following variables: surface pressure, temperature, humidity, wind, rainfall, snowfall, and cloudiness; upper-air temperature, pressure–altitude, wind, and humidity; ocean temperature, salinity, and current; the location and depth of land ice, sea ice, and snow; the surface insolation, ground temperature, ground moisture, and runoff. This inventory should identify the length of the observational record, the data quality, and the state of its availability. In addition to the usual data sources, efforts should be made to locate data from private sources, older records, and unpublished climatological summaries. Although some of these data have been summarized, no overall inventory of this type exists.

Selected portions of these data should be systematically transferred to *suitable computer storage,* in a format permitting easy access and screening by variable, time period, and location. These data should then be used to compute in a systematic fashion a basic set of *climatic statistics* for as many time periods and for as many regions of the world as possible. These should include the means, the variances, and the extremes for monthly, seasonal, annual, and decadal periods, for both individual stations and for various ensembles of stations up to and including the entire globe. Research should also be devoted to the effects of instrumental errors, observational coverage, and analysis procedures on climatic statistics.

Recognizing that these data have large differences in quality, coverage, and length-of-record and were often collected as by-products of other studies, new four-dimensional *climatological data-analysis schemes* should be developed, based on suitable analysis methods or models, to synthesize as much of the missing information as possible while making maximum use of the available data. Efforts should also be devoted to the design of suitable computerized graphical display and output.

Once such syntheses are available, we recommend that suitable *climatological diagnostic studies* be made using dynamical climate models to generate systematically the various auxiliary and unobserved climatic variables, such as evaporation, sensible heat flux, surface wind stress, and the balances of surface heat, moisture, and momentum. Such data, of course, would be artificial but may nevertheless be of diagnostic use. Insofar as possible, the pertinent statistics of the atmospheric and oceanic general circulations and their energy, momentum, and heat balances should be determined.

The results of such analyses should be made available in the form of new *climatological atlases,* supplementing and extending those now available for scattered portions of the record and for selected regions of the world. The widely used climatological summaries of Sverdrup (1942), Möller (1951), and Budyko (1963), for example, are largely based on the subjective analyses of older data of uncertain quality. Other analyses are more authoritative (Oort and Rasmusson, 1971; Newell et al., 1972) but are in need of extension.

We wish to emphasize the great importance of the potentially unmatched coverage of observations from satellites. Those that are of climatic value should be systematically cataloged and summarized and made available on as timely a basis as possible. These should include observations of cloud cover, snow, and ice extent; planetary albedo; and the net radiation balance. As remote techniques for measuring the atmosphere's composition, motion, and temperature structure (and the surface temperature of land and ocean) are developed, these data should be systematically added to the climatological inventory. They should also be used in the analysis and model-based diagnostic efforts described above and in the climatic index monitoring program outlined below. The presently available summaries of such data (e.g., Vonder Haar and Suomi, 1971) have yielded important new results and should be continued on an expanded basis.

Historical Data As noted in Appendix A, a wealth of information has been recorded on past variations of weather and climate in historical sources such as books, manuscripts, logs, and journals during the past several centuries. While much of these data are fragmentary and not of a quality comparable with that of instrumental observations, it is nevertheless of value. We therefore recommend that

An organized effort be made to *locate, classify, and summarize historical climatic information* and to identify and exploit new sources. From the studies of this sort that have already been made (e.g., Bryson and Julian, 1963; LeRoy Ladurie, 1971; Lamb, 1968, 1972), it is clear that these efforts should involve historians, archeologists, and geographers on an international scale.

Efforts be made to relate this material to data from other proxy sources whenever possible, and efforts made to interpret and focus the material in a climatologically meaningful way.

Proxy Data We recognize the unique value of proxy data for studies of climatic change. Such data are obtained from the analysis of tree-ring growth patterns, glacier movements, lake and deep-sea sediments, ice cores, and studies of soil and periglacial stratigraphy. Data from tree rings, annually layered lake sediments, and some ice cores are capable of providing information for individual years, while those from other sources provide more generalized climatic information on time scales of decades, centuries, and millenia. Such data constitute the only source of records for the study of the structure and characteristics fluctuations of ancient climates. As discussed in Appendix A of this report, some of these past climates were quite different from the present regime and provide our only documentation of the extreme states of which the earth's climatic system is capable.

Because all proxy climatic data may contain both bias and random error components, it is essential that a variety of independent proxy records be studied. It is important that coverage be as nearly global as possible, since most of the information on climatic variations is contained in the spatial patterns of the data fields. While noting that some such activity is already in progress, we urge that the assembly and analysis of paleoclimatic data be initially focused on four time spans (see below). This represents a strategic decision in order to make the best use of the available resources. In each area it is important that steps be taken to increase greatly the degree of coordination and cooperation within the paleoclimatological community, and that the cross-checking of overlapping data sets, the development of complementary and independent proxy data sources, and the calibration against instrumentally observed data be undertaken whenever possible.

The last 10,000 years. This is the interval within which we may hope to gain insight into the current interglacial period by the systematic assembly of a wide variety of proxy climatic data. This is also the interval of greatest practical importance for the immediate future. For this period particular attention should be given to six techniques:

Studies of the structural, isotopic, and chemical properties of *tree rings* should be intensified and extended to a global coverage. Since forests cover large areas of the globe, it is possible in principle to develop climatic records over extensive continental areas and to reconstruct the spatial patterns of past climate for the past several centuries or millenia. The amount of effort depends on the availability of suitable trees and on the resolution required in the climatic reconstruction. Data on variations of the density of wood from x-ray techniques and on the concentrations of trace elements and of stable isotopes of carbon, hydrogen, and oxygen in well-dated rings should also be developed. Special efforts should be made to calibrate the few millenia-long tree-ring records with information from other suitable proxy data sources, such as pollen, varves, and ice cores.

Studies of *pollen records* in lakes and bogs should be extended. Most pollen analyses to date have concerned bogs created by the retreat of the last continental ice sheet. In order to permit synoptic reconstruction of the global vegetational record for the past 10,000 years or so, pollen analyses with extensive ^{14}C dating should be extended to the nonglaciated areas of the world, particularly to low-latitude regions and to the southern hemisphere.

Studies of the *polar ice caps* should be expanded. This should include additional short ice cores in widely distributed locations, in both Greenland and Antarctica, and more detailed isotopic analyses.

Studies of the major *mountain glaciers* should be expanded, to obtain additional information on the various glaciers' advances and retreats, using chronological control where possible.

Studies of *ocean sediments* in the few basins of known high deposition rates should be greatly expanded. Particularly near the continental margins, the synoptic reconstruction of even the decadal variations of sea-surface temperatures (and possibly of currents as well) would be of great paleoclimatic interest. This effort will involve lithologic, faunal, and isotopic analyses of long cores collected specifically for this purpose.

The records from *varved sediments* in closed basin lakes or landlocked seas should be extended. Such data are particularly sensitive to the climatic fluctuations in arid regions and would further our knowledge of the long-term behavior of deserts and drought.

The last 30,000 years. This interval is dominated by the waxing and waning of continental ice sheets. In this interval the radiocarbon dating method provides a good chronology, and the possibilities for studying the relative phases of different proxy climatic records on a global basis are a maximum. In this period particular efforts should be made in the following areas:

Pollen records for the interval 10,000 to 30,000 years ago should be obtained in a wide variety of sites in both hemispheres.

Ice-margin data should continue to be collected for northern hemisphere glaciers and should be extended into southern hemisphere mountain areas.

Additional *deep-sea cores* should be obtained, especially in the Pacific and Southern Oceans, in order to reveal further the geographic pattern of marine paleoclimates. These data would be particularly useful from high-deposition-rate basins.

Additional data should be obtained on the fluctuations in the extent and volume of the polar *ice sheets* during this time interval. Particular attention should be given to the smaller ice sheets, such as the West Antarctic and Greenland ice sheets, which react more rapidly to climatic variations.

More extensive analyses of *sea-level records* should be made, emphasizing the removal of tectonic and isostatic effects. Present studies on raised coral reefs should be extended, and estuarine borings should be carefully dated and given thorough lithologic analysis.

The last 150,000 years. Here we should seek to increase our knowledge of the last 100,000-year glacial–interglacial cycle. This interval includes the last period in the climatic history of the earth that was evidently most like that of today. The data of this period also provide the best example of how the last interglacial period ended. Efforts should be made to further develop a number of proxy data sources, including:

Extensive collection and analyses of *marine sediment cores* to provide adequate global coverage of the world ocean.

Further studies of the fluctuations of the *Antarctic and Greenland ice caps,* with emphasis on records extending beyond the beginning of the last interglacial. This should include a geographic network of ice cores of sufficient length to penetrate this time range, of which those at Camp Century, Byrd, and Vostok are now the only examples.

Further systematic study of the *loess-soil sequences* in suitable regions around the world, including Argentina, Australia, China, and the Great Plains of North America.

Systematic studies of *desert regions* and arid intermountain basin areas in order to examine the patterns of long-term changes in aridity. Present records are limited to about the last 40,000 years, and their extension will require long borings in selected lakes and playas.

Extended studies of *sea-level variations* from coral reef and island shorelines features.

Further studies of long pollen records covering previous interglacial periods. This should include data from previously unsampled regions of the world, particularly in the southern hemisphere.

The last 1,000,000 years and beyond. Fluctuations in this time range should not be ignored simply because of their antiquity. Here we have the opportunity to compare the circulation patterns that have characterized the last several full-glacial and interglacial periods, and thereby to contribute evidence on the question of the degree of determinism of the earth's climatic system. Efforts should therefore be made to extend suitable proxy records into this time range, including:

Additional *marine sediment cores* of sufficient length (say, up to 100 m long) to cover several glacial cycles should be obtained. This will require new innovations in drilling technology, as piston cores do not penetrate deeply enough for this purpose, and rotary drills presently in use greatly disturb the sedimentary record.

The record of the *Antarctic ice sheet* (and the associated sea-level variations) should be extended as far back as possible and in as much detail as possible. This ice mass is a living climatic fossil and may contain information about the global climate for the past several million years.

Climatic Index Identification and Monitoring In addition to the data provided by conventional surface and upper-air observations, climatic studies require other contemporary data that are not now readily available. The one hope for obtaining truly global coverage of many current climatic variables rests with satellite observations. We expect that climatic studies in the foreseeable future will have to rely on a combination of conventional observations, satellite observations, and special observations designed to monitor selected climatic variables as discussed below. We should therefore make full use of the temporary expansion of the observational network planned for the FGGE in 1978 in order to design a longer-lived climatic observing program. In addition, efforts should be made to process the monitored data from both satellites and other systems into forms that are useful for climatic studies. Support should be given to the development of new satellite-based observational techniques, including those designed to monitor the oceans and the earth's surface.

There remain, however, a number of processes that are important to climate that are now beyond the reach of satellite observations. Primary among these is the pattern of the planetary thermal forcing, which drives the atmospheric and oceanic circulation, and the related balance of energy at the earth's surface. Even a measurement of the average pole-to-equator temperature difference tells us something about the circulation; and, in a similar way, the discharge of a river gives us some information on the hydrologic balance in the river's basin.

Such measurements, which represent time and space integrals of climatically important procesess, we term "climatic indices." While efforts to monitor indices of this sort are already under way, we recommend that further efforts be made to identify and monitor a variety of such indices in a coordinated and sustained fashion, as part of a comprehensive global Climatic Index Monitoring Program (CIMP) whose elements are outlined below.

Atmospheric Indices The heat balance of the atmosphere is basic to the character of the general circulation and hence is a principal determinant of climate. It is therefore important that the primary elements of this balance be monitored with as much accuracy and with as nearly global coverage as possible. In particular, we recommend that further efforts be made to

Monitor the solar constant and the spectral distribution of solar radiation with appropriate satelliteborne instrumentation.

Monitor the net outgoing shortwave and long-wave radiation by satellite-based measurements, from which determinations of the absorbed radiation and planetary albedo may be made.

Monitor the latent heat released in large-scale tropical convection, possibly with the aid of satellite cloud observations.

Develop methods to monitor remotely the surface latent heat flux into the atmosphere, possibly with the aid of satellite measurements of the vertical distribution and total amount of water vapor. These methods (and those for the sensible heat flux discussed below) will require calibration against field appropriate measurements, especially over the oceans.

Develop methods to monitor remotely the surface sensible heat flux into the atmosphere, especially that from the oceans, such as occurs in winter off the east coasts of the continents and in the higher latitudes. Efforts should also be made to monitor remotely the vertical sensible heat flux that occurs as a result of convective motions both over the oceans and over land.

Expand the satellite monitoring of global cloud cover to include information on the clouds' height, thickness, and liquid water content, so that their role in the heat balance may be determined.

Monitor the distribution of surface wind over the oceans, possibly by radar measurements of the scattering by surface waves or from the microwave emissivity changes created by foam.

Oceanic Indices In view of the fundamental role the oceans play in the processes of climatic change, special efforts should be made to monitor those oceanic variables associated with large-scale thermal interaction with the atmosphere. In addition to the low-level air temperature, moisture, cloudiness, surface wind, and surface radiation, the surface heat exchange depends critically on the sea-surface temperature and heat storage in the oceanic surface layer itself. We therefore recommend that further efforts be made to

Monitor the worldwide distribution of sea-surface temperature by a combination of all available ship, buoy, coastal, and satellite-based measurements. Sea-surface temperature analyses, such as now performed operationally by the Navy's Fleet Numerical Weather Central in Monterey, should be extended and supplemented for climatic purposes on a global basis by improved satellite observations capable of penetrating cloud layers. The drifting buoy observations of sea-surface temperature planned for the FGGE should be expanded and maintained on a routine basis.

Monitor the heat storage in the surface layer of the ocean by a program of observations from satellite-interrogated expandable drifting buoys and by expendable bathythermograph (XBT) observations from ships-of-opportunity in those areas of the world ocean traveled by commercial ships. It is estimated that there are several hundred such transits each year across most major oceans of the world. An expansion of XBT observations from merchant ships-of-opportunity is being undertaken by the North Pacific project (NORPAX), in cooperation with the Navy's Fleet Numerical Weather Central and NOAA's National Marine Fisheries Service. Similar programs should be undertaken in the other oceans, and especially in the oceans of the southern hemisphere, with special efforts made to place instruments aboard ships on unconventional routes and on selected government vessels. This XBT program should be supplemented by buoy measurements in selected locations

A NATIONAL CLIMATIC RESEARCH PROGRAM 75

and by XBT's launched from aircraft on meridional flight paths in the more inaccessible ocean areas.

Expand the present data buoy programs now under way by NOAA and others, so that the volume and heat transport of the major ocean currents can be monitored. Suitably deployed bottom-mounted sensors, moored buoys, or both should be used to monitor the transport of the Gulf Stream, Kuroshio, and Antarctic circumpolar currents in selected locations, such as is planned for the Drake Passage as part of the International Southern Ocean Studies (ISOS). The water mass balance of individual basins such as the Arctic should also be monitored.

Monitor the complete temperature structure in selected regions of the ocean, such as meridional cross sections through the major gyral circulations. The several long-term local observational series (such as the Panulirus, Plymouth, and Murmansk sections) should be maintained and new efforts started in regions of special interest.

Monitor the vertical salinity structure of the oceans in those high-latitude regions where salinity plays an important role in determining the density field of the upper ocean layers. Near-surface salinity is also important in regions where ocean bottom water is formed, such as in the Weddell Sea. This might best be done by a combination of moored buoys and ship observations.

Monitor the large-scale distribution of sea level by the use of an expanded network of tide gauges. Such a measurement program at island sites in the equatorial Pacific is being undertaken in connection with NORPAX, and other measurements are planned in the Indian Ocean as part of the Indian Ocean Experiment (INDEX). Radar altimeters such as those proposed for the SEASAT-A satellite should also be useful for this purpose.

Monitor the oceanic chemical composition at selected sites and in selected sections, including the concentrations of dissolved gases and trace substances. Such measurements now being performed as part of the GEOSECS program should be expanded and continued.

Cryospheric Indices In view of the great influence of snow and ice cover on the surface energy balance, further efforts should be made to

Monitor the distribution of sea ice in the polar oceans and the ice in major lakes and estuaries. Efforts should also be made to measure as many as possible of the ice's physical properties by remote sensing.

Devote further study to the current mass budgets of the Antarctic and Greenland ice caps, from both glaciological field observations and from airborne and satellite measurements. Such observations should include changes in ice-edge locations, in the numbers and sizes of icebergs, and in the ice caps' firnline height. Methods for the remote aerial sensing of surface temperature and possibly ice accumulation rate should also be further developed.

Extend the monitoring of the movement and mass budget of selected mountain glaciers.

Monitor the extent, depth, and characteristics of worldwide snow cover.

Surface and Hydrologic Indices In association with the monitoring of the elements of the surface heat balance, and of the various oceanic and cryospheric climatic indices, initially lower priority but nevertheless important efforts should be made to

Monitor the natural changes of surface vegetative cover, possibly by observations from earth resources satellites.

Monitor the variations of soil moisture and groundwater, possibly by satellite-based techniques.

Monitor the flow and discharge of the major river systems of the world.

Monitor the level and water balance of the major lakes of the world.

Monitor the total precipitation (especially rainfall over the oceans), possibly by satelliteborne radar observations and surface gauges.

Composition and Turbidity Indices In view of the role that atmospheric constituents and aerosols play in the heat balance of the atmosphere, further efforts should be made to

Monitor the chemical composition of the atmosphere at a number of sites throughout the world, with particular reference to the content of CO_2. Measurements such as those at Mauna Loa should be continued and extended to additional selected sites. The composition of the higher atmosphere should also be periodically determined, especially the water vapor in the stratosphere and the ozone concentration in the stratosphere and mesosphere.

Monitor the total aerosol and dust loading of the atmosphere, together with determinations of the vertical and horizontal aerosol distribution, by an extension of such programs as NCAR's Global Atmospheric Aerosol Study (GAARS). In addition to turbidity measurements, the aerosol particle-size distribution and optical properties should be determined when possible. Efforts should also be made to monitor the

A NATIONAL CLIMATIC RESEARCH PROGRAM 77

occurrence of large-scale forest fires and volcanic eruptions, together with estimates of their particulate loading of the atmosphere.

Anthropogenic Indices In view of man's increasing interference with the environment, further efforts should be made to

Monitor the addition of waste heat into the atmosphere and ocean. Although the present levels of thermal pollution are relatively small on a global basis, steadily increasing levels of energy generation pose a threat to the stability of at least the local climate and possibly the larger-scale climate as well. Therefore both the local thermal discharges of power generating and industrial facilities should be monitored, along with the thermal pollution from urbanized areas.

Monitor the climate-sensitive chemical pollution of the atmosphere and ocean. Measurement programs such as those of the Environmental Protection Agency and the Atomic Energy Commission should be expanded on a global basis and extended to the oceans.

Monitor the changes of large-scale land use, including forest clearing, irrigation, and urbanization, possibly by the use of earth resources satellites.

Summary of Climatic Index Monitoring A summary of the elements of the recommended program is given in Table 6.1. Here we have not made an assessment of the required accuracy of the various monitored indices, nor has the capability of presently available instrumentation been thoroughly reviewed. Further analysis is also needed to determine the characteristic variability of each climatic index. In general, the surface heat and hydrologic balances should be monitored with an accuracy of a few percent, so that space- and time-averaged climatic statistics will have at least a 5 percent accuracy. It is important that this monitoring activity be undertaken on a continuing and long-term basis for at least two decades in order to assemble a meaningful body of data for climatic analyses. As noted below, these efforts should be coordinated on an international scale and be a part of an international climatic program.

Research Needed on Climatic Variation

We here outline the research that we believe needs to be performed, in terms of model development, theoretical research, and empirical and diagnostic studies. While research in some of these areas is already under way as part of GARP activities in anticipation of the FGGE, these

TABLE 6.1 Summary of Climatic Index Monitoring Program (CIMP)

Variable or Index	Method	Coverage	Effort Required [a]	Frequency Required [b]
Atmospheric indices				
Solar constant	Satellite	Global	N	W
Absorbed radiation, albedo	Satellite	Global	P	W
Latent heating	Satellite	Global	N	W
Surface latent heat flux	Satellite	World ocean	N	W
Surface sensible heat flux	Satellite	Regional	N	W
Cloudiness	Satellite	Global	P	W
Surface wind over ocean	Radar scattering	World ocean	N	W
Oceanic indices				
Sea-surface temperature	Ships, satellites, buoys	World ocean	E	W
Surface-layer heat storage	XBT, AXBT, buoys	Mid-latitude and low-latitude oceans	E, N	W
Heat transport	Moored buoys	Selected sections	N	W
Temperature structure	Ships	Selected sections	E	S
Surface salinity	Ships, buoys	High latitudes	E	W
Sea level	Tide gauges	Selected coastal and island sites	E	W

Floating ice extent	Satellite	Polar seas, lakes	E	M
Ice-sheet budget parameters	Satellite	Greenland, Antarctica	N	Y
Mountain glacier extent	Satellite	Selected sites	E	Y
Snow cover	Satellite	Continents	E	M
Surface and hydrologic indices				
River discharge	Flow gauges	Selected sites	E, N	W
Soil moisture	Satellite	Land areas	E	W
Lake levels	Gauges	Selected sites	E	W
Precipitation	Satellite, radar, gauges	Global	E	W
Composition and turbidity indices				
Chemical composition	Sampling	Selected sites	E	S
Aerosols and dust	Satellite	Global	E	W
Anthropogenic indices				
Thermal pollution	Sampling	Continents and coasts	N	W
Air and water pollution	Sampling	Global	E	W
Land use	Satellite	Continents	E	Y

[a] N, completely new monitoring effort required; E, expansion of present monitoring efforts required; P, present (or slightly expanded) monitoring efforts satisfactory but coordination and further analysis required.
[b] W, weekly (or possibly daily in some cases); M, monthly; S, seasonally; Y, yearly (or possibly decadal in some cases).

efforts are the necessary ingredients of the much broader climatic research program that we recommend be carried out in the years ahead.

Theoretical Studies of Climatic Change Mechanisms We recognize the importance of theoretical studies in a problem as complex as climatic variation and the essential interaction that must take place between theory and the complementary observational and numerical modeling studies. Our present knowledge of the mechanisms of climatic variation is so meager, however, and progress in this area so difficult to anticipate, that any recommendations are subject to modification as new avenues of attack open up or as old ones prove fruitless. There are, however, certain fundamental problems to which further study must be directed:

The question of the degree of *predictability of (natural) climatic change* must be given further theoretical attention. While the local details of weather do not appear to be predictable beyond a few weeks' time, the consequences of this fact for climatic variations are not clear. In such studies a clear definition of the internal climatic system needs to be made, and particular attention must be given to the roles of the ocean and ice. This question has an obvious and important bearing on our eventual capability to predict climatic variation.

The related question of the possible *intransitivity* of climatic states needs further study, again with particular attention to the oceans and ice. The whole question of climatic variation may be viewed as a stability problem for a system containing elements with very different time constants, and support should be given to such theoretical approaches.

Theoretical research should be directed to the nature and stability of the various *climatic feedback mechanisms* identified earlier, particularly those involving the sea-surface temperature, cloudiness, albedo, and land-surface character.

Further theoretical research should be directed to the general problem of the development of *statistical–hydrodynamical* representations of climate and to the parameterization of transient phenomena on a variety of time and space scales.

Additional theoretical studies should also be made of specific climatic phenomena, such as drought and the growth of arid regions, ice ages and the stability of polar ice cover, and the effects of global pollution from natural and artificial sources.

Atmospheric General Circulation Models The global dynamical models of the atmospheric general circulation (or GCM's) that have been developed in recent years represent the most sophisticated mathematical tools ever available for the study of this system and are the testing grounds for many of our theoretical ideas. The latest versions of these models (see Appendix B) embody much of the physics that governs the larger scales of atmospheric behavior, along with physical parameterizations of smaller-scale processes. In addition to the simulation of the free-air temperature, pressure, wind, and humidity distributions over the globe with a resolution of several hundred kilometers, such atmospheric models provide solutions for the various components of the heat and moisture balances, such as the fluxes of shortwave and long-wave radiation, sensible heat flux, evaporation, precipitation, surface runoff, and ground temperature. The surface boundary conditions usually assumed are the distributions of sea-surface temperature and sea ice, and the assumption of a heat balance over land surfaces. After a spin-up period of a month or so during which the temperature comes into statistical equilibrium (with the sun's heating and the ocean surface temperature), the average global climate simulated by such models shows a reasonable resemblance to observation; several examples of such simulations are shown in Appendix B.

In order to improve the fidelity of such global atmospheric models for the simulation of the various processes of climatic change, and to ensure their increased availability for the conduct of climatic experiments, efforts should be made to

Improve the models' treatment of *clouds,* especially those of the nonprecipitating high-altitude cirrus and the low-level stratus type. Account should be taken of the liquid-water content of clouds and the full interaction of clouds with atmospheric radiative transfer. Attention should also be given to the modeling of cloud evaporation and advection.

Improve the parameterization of *turbulent, convective, and mesoscale processes* by comparing the performance of alternative schemes against appropriate observations of the fluxes of heat, moisture, and momentum. Particular attention should be given to improved parameterizations of the fluxes within the surface boundary layer, to the parameterization of cumulus convection, and to the treatment of energy flux by gravity waves.

Improve the treatment of *ground cover* and land usage in the calculation of the surface heat and moisture balances. Particular attention should be given to the improvement of the prognostic schemes for *snow cover,* as this may prove of importance in seasonal climatic variations.

Parameterize the role of *aerosols* in such models, so that the effects of both natural and anthropogenic particulates on the heating rate of the atmosphere may be determined.

Improve the *numerical resolution* of the solutions by the use of finer grids (or the use of graded meshes in regions of special interest) and increase the *computational efficiency* by the development of more accurate numerical algorithms and improved solution methods.

Simulate the *annual cycle* of atmospheric circulation with models using observed forcing functions to obtain the surface fluxes of heat, momentum, and water vapor. Such numerical integrations are necessary in order to ensure adequate model calibration and to simulate climatic statistics for the atmosphere.

Determine the *noise level* or *sensitivity* of the model-simulated climate to changes in the initial conditions (including random errors) and to changes in the parameterizations of the model. Such studies are necessary in order to determine the physical significance of numerical climate-change experiments made with atmospheric models.

Oceanic General Circulation Models The oceanic general circulation models (or GCM's) are generally at a less advanced stage of development than their atmospheric counterparts and have only recently been extended to the global ocean (see Appendix B). With the surface boundary conditions of specified thermal forcing and wind stress (plus the kinematic and insulated wall boundary conditions at the bottom and lateral sides of the ocean basin), such models simulate with fair accuracy the large-scale distributions of ocean temperature and current with a resolution of several hundred kilometers. If the density structure is specified from observations, a model will spin up from rest in a few months' time and show a reasonable correspondence with observed drift current patterns at the surface. The simulated transport of the major western boundary currents in the models is generally less than that indicated by available observations but nevertheless quantitatively more accurate than the predictions of previous theories. Areas of coastal and equatorial upwelling show the same strong relationship to the surface wind-stress pattern in the model as is observed in the real ocean.

The more relevant calculation with respect to climate modeling is one in which the density field as well as the velocity field is predicted from boundary conditions that determine the vertical flux of momentum, heat, and water at the ocean surface. However, this problem involves much longer time scales—the spin-up time of a prestratified ocean is of the order of two or three decades; but if changes in the abyssal thermal structure are to be predicted, then the "turnover" time of the ocean is the order of several centuries. Preliminary results (see, for example, Bryan and Cox, 1968) show that such models can successfully simulate the gross features of the density structure of the world ocean, although more detailed calculations must be made to provide a critical test.

In order to improve the accuracy of ocean models and to lay the foundation for their successful coupling with atmospheric models, efforts should be made to

Improve our knowledge of the structure, behavior, and role of *mesoscale eddies* in the ocean. In the atmosphere there is a peak in the kinetic energy spectrum observed at wavelengths of a few thousand kilometers, whereas in the ocean the peak kinetic energy is in eddies that have a radius (or quarter-wavelength) of the order of 10^2 km. Thus an ocean circulation model requires about an order of magnitude greater horizontal resolution to resolve its most energetic eddies than does an atmospheric GCM. Further field studies such as those conducted under the Mid-ocean Dynamics Experiment (MODE), the North Pacific Experiment (NORPAX), and those planned under the joint Soviet–American (POLYMODE) experiment, are needed to determine the transfer of heat and momentum by such eddies. Such observational experiments should provide the basis for the interpretation of high-resolution numerical experiments, which are necessary to resolve the details of the eddy motions and to establish their role in the oceanic general circulation.

Intensify research on the *parameterization* of turbulent and mesoscale motions both in the surface mixed layer and the deeper ocean layers, including thermohaline convection, so that the results of field measurements may be usefully incorporated into global ocean circulation models.

Improve the prediction of sea-surface temperature and heat transport by the inclusion of the depth and structure of the *surface mixed layer* as a predicted variable in oceanic general circulation models. This should include experiments on the *numerical forecasting* of the oceanic surface layer, as driven by observed surface conditions, and the formation and behavior of pools of anomalously warm or cold water.

Simulate the *annual cycle* of sea-surface temperature and currents with models using observed forcing functions to obtain the surface fluxes of momentum, heat, and water (precipitation minus evaporation). Such numerical integrations must be carried out over several annual cycles, in order to ensure adequate model calibration and to simulate climatic statistics for the ocean.

Subject the ocean models to the same kind of *diagnostic testing and sensitivity analysis* as performed for atmospheric models, in order to determine the roles of possible oceanic feedback processes and the levels of predictability associated with various oceanic variables.

Apply high-resolution versions of global oceanic circulation models (or regional versions thereof) to the study of the behavior of local intense currents, such as the eddying motion of western boundary currents and the structure of equatorial currents.

Develop more accurate models of *sea ice*, which include the effects of salinity and the dynamic and thermodynamic factors governing the distribution of the polar ice packs. The data base being assembled by the Arctic Ice Dynamics Joint Experiment (AIDJEX) in the Beaufort Sea should be useful in the design of models that can predict those properties of the polar ice pack that are important in the surface heat balance, such as the ice thickness and the occurrence of open-water leads.

Search for new *computational algorithms* for predicting oceanic circulation that will provide the greatest accuracy for the least possible cost. At present, the methods used in modeling the ocean are similar to those used in the atmospheric GCM's. The presence of lateral boundaries and the need to resolve mesoscale motions may make alternative numerical methods of particular use in numerical ocean models.

Coupled Global Atmosphere–Ocean Models   Tests of climatic change extending over one or more years are not adequate unless they are made with a model of the coupled ocean–atmosphere system. While the uncoupled atmospheric and oceanic GCM's are useful for many purposes, the thermal and mechanical coupling between the ocean and atmosphere is fundamental to climatic variation. We note that a global ocean model may require only a fraction of the computational effort needed by an atmospheric GCM of the same resolution but emphasize that care must be taken to avoid erroneous drift in the simulated climate due to systematic biases in the model or in the oceanic initial state.

Assuming that coupled models (CGCM's) will incorporate the developments and improvements recommended above for the separate atmospheric and oceanic models, emphasis should be given to the following research with CGCM's:

Investigation of the *simulated climatic variability*, on seasonal and annual time scales, of all climatic variables of the coupled system, including the simulated exchange processes at the air–sea interface.

Of particular importance in the coupled models is the simulation of the sea-surface temperature, as this has a key role in the evolution of the system. This will require integration over many years of simulated time in order to generate adequate climatic statistics and to examine the models' stability. Particular attention should be given to evidence of climatic trends and intransitivity in the numerical solutions. The statistics of such simulations with CGCM's will also prove valuable in the calibration of statistical climate models.

The *sensitivity* of the climate simulated by coupled models should be systematically examined in experiments extending at least through an annual cycle. These studies should include the climatic consequences of uncertainties in the simulations' initial state (including random errors), in the parameterization of the various physical processes (such as convection, cloudiness, boundary-layer fluxes, and mesoscale oceanic eddies), and in the computational procedures. Such studies are necessary in order to establish the characteristic noise levels of the models and are of great importance in the use of the models for climate experiments.

A program of *climate change hypothesis testing* should be undertaken with coupled models, as soon as their stability and calibration are reasonably assured. This should include examination of the various feedback mechanisms among components of the climatic system, such as ice and snow, cloudiness, sea-surface temperature, albedo, radiation, and convection.

The coupled models should be used in a program of *long-range integrations* with observed initial and boundary conditions, in order to assess both their overall fidelity and their usefulness as long-range or climatic forecasting tools.

Although not a research task in itself, special efforts should be made to appropriately store, analyze, and display the rather staggering amounts of data generated during the integration of CGCM's, so that subsequent diagnosis can be performed efficiently.

Statistical–Dynamical Climate Models Although the coupled numerical models of the global circulation offer the most comprehensive and detailed solutions available, even with the fastest computers envisaged relatively few century-long climatic simulations will be possible, and it is likely that none will be performed for periods as long as a millenium. Such models will therefore find their greatest use in climatic research in the exploration of the character of relatively short-period (say annual to decadal) climatic variations and in the calibration of other, less-detailed models. We therefore emphasize that statistical–dynamical climate models (defined as those in which the structure and motion of the individual large-scale transient disturbances are not resolved in detail) will have to be used to simulate the longer-period climatic variations. While such models provide less resolution of the details of climatic change, they may display less climatic noise than do the global circulation models.

In order to ensure the availability of the hierarchy of models needed in a comprehensive research program on climatic change, the following research should be carried out:

Statistical–dynamical models of the coupled time-dependent atmospheric and oceanic circulation should be constructed and calibrated that embody suitable *time- and space-averaged representations* of the climatic elements. In their extreme form, such models address the steady-state globally averaged quantities, while others, for example, consider time-dependent zonally averaged variables. Further efforts should be made to represent the climatically important land–sea distribution in such models and to calibrate them systematically against observations as well as against other climatic models.

Simulation of climatic variation over extended time periods should be made by the integration of suitably calibrated time-dependent statistical–dynamical models. Depending on the time range, appropriate components of the climatic system's atmosphere, hydrosphere, cryosphere, lithosphere, and biosphere should be introduced, along with appropriate variations of the external boundary conditions (see Figures 3.1 and 3.2).

Coupled time-dependent models in which the global circulation is represented by *low-order spatial resolution* should also be further developed, such as those using a limited number of orthogonal components or spectral modes.

Coupled models should be constructed and calibrated that embody new forms of *time-averaged representations* of the climatic system. We recognize that the parameterization of the effects of the transient eddies poses a difficult problem in statistical hydrodynamics and urge that full use be made of both model-generated and observed statistics, as well as of theory, to develop a variety of such models for different types and ranges of time averaging.

In each type of statistical–dynamical model, particular attention should be given to the *inclusion of the ocean and ice*. In such models, attention should also be given to the possibility of treating the atmosphere statistically while simulating the ocean in detail and perhaps of treating both the atmosphere and ocean statistically while simulating the growth of ice sheets in detail. It is particularly important that such models be *calibrated* with respect to both the mean and variance of the climatic elements and that their *stability and sensitivity* be systematically determined.

Empirical and Diagnostic Studies of Climatic Variation Although we have recommended some diagnostic and empirical studies in connection with the analysis of instrumental and proxy climatic data, such studies should also be made on a phenomenological basis as part of the climatic analysis and research program. As the record of past climates is made more complete, there will be increased opportunity to carry out such investigations with both instrumental and proxy data. In particular:

Studies should be made of the *temporal and spatial correlations* among various data, including regional and global estimates of the trends of key climatic elements such as temperature and precipitation.

Further empirical studies should be made of the surface oceanic variables of temperature, salinity, sea level, and sea ice and of the planetary heat balance, albedo, and cloudiness from satellite-based observations. The studies of Bjerknes (1969), Kukla and Kukla (1974), Namias (1972a), and Wyrtki (1973) are examples of the sort of *empirical synthesis* that can be achieved and should be systematically extended to other regions of the world and to other climatic variables. In these efforts, particular attention should be given to the various possible climatic feedback processes and to the forcing functions of the general circulation. Here the diagnostic use of climatic models should prove valuable.

Further studies should be made of the *statistical characteristics* of climatic data, both observed and simulated. Power spectrum analyses should be made for as many variables and locations as possible, and with the longest records available, as the spectrum's "redness" has an important bearing on questions of climatic cycles and climate prediction.

Needed Applications of Climatic Studies

Although closely related to the climatic data analysis and climatic research recommended above, the needed applications of climatic studies (and of climate models in particular) are so important that they warrant identification as a separate component of the program. It is in these applications that the program reaches its fruition, and if attention to them is delayed until our understanding is complete or our models perfect, they may never be undertaken. With due regard for scientific caution, we believe that the time has come for a vigorous attack on the areas of climate model application described below.

Simulation of the Earth's Climatic History

The evidence presented in Appendix A (and summarized in Chapter 4) shows that the climatic history of the earth has been remarkably variable and that this history provides information that is of value in the study of present and possible future climates. The data assembled by paleoclimatologists show conclusively that the flora, fauna, and surface characteristics of many regions of the world have often been markedly different in past times than they are today. Compared with this long-period panorama, instrumental observations provide a frustratingly short record.

It is at this juncture that the intersection of paleoclimatic and numerical modeling studies offers the most promise: the global climatic models have the potential ability to simulate at least a near-equilibrium approximation to past climates subject to the appropriate geological boundary conditions, while the paleoclimatic records can be used as verification data. Initial efforts in this direction have already begun (see Chapter 5), and we may expect increasing insight into the nature of past climates as both the models and proxy data base improve.

In order to explore the nature of past climates systematically and to lay the foundation for the study of possible future climates, the following studies should be made:

The *geophysical boundary conditions* at a number of selected times in the history of the earth should be systematically assembled with a view toward their use in climate models. This should include global data on the continental land-mass positions and elevations, sea-level ice-sheet elevations and margins, sea-ice extent, soil type and vegetative cover, and surface albedo. Estimates should also be made of the earth's rotation rate and of the solar insolation (due to orbital parameter changes). The selection of the time period might be based on criteria such as the occurrence of an ice age, the distribution of the continents and mountains, the opening or closing of a major oceanic passage, or the large-scale flooding or draining of lowlands. Periods of particular climatic stress such as indicated by the disappearance of species might also be considered.

The various *proxy records* of temperature, salinity, and precipitation should also be systematically assembled for the same selected times, to serve as verification data for the coupled climate models' simulations and as possible input or boundary conditions for uncoupled models.

Dynamical global models should be used to simulate the *quasi-equilibrium paleoclimate* at selected times in the past when the boundary conditions external to the ocean–atmosphere system can be reasonably well specified. Such experiments should be focused on times when the global climate might be expected to be in a particularly interesting state (as judged from the available geological and proxy evidence) or when the climate might be expected to be in the process of changing most rapidly from one characteristic regime to another. The simulations should extend long enough to accumulate realistic climatic statistics and should use the assembled paleoclimatic data for vertification. By using part of the paleoclimatic evidence (namely, the sea-surface temperature) as a boundary condition, atmospheric GCM's may also be used for this purpose.

Coupled statistical–dynamical models, or other coupled climate models, should be used to simulate the *time-dependent climatic evolution* between the various "equilibrium" states identified above. For this application the dynamics of *ice sheets* should be incorporated into the coupled ocean–atmosphere models and note taken of the possible time dependence of the remaining boundary conditions, such as solar radiation and continental drift. In particular, the astronomical changes of seasonal radiation resulting from the variation of the earth's orbital parameters should be incorporated in a climate model, and the resulting simulated climatic changes compared with the paleoclimatic evidence. This recommendation parallels one made earlier in connection with the development of the statistical–dynamical models themselves.

Studies should be made of possible methods to accelerate the simulation of quasi-equilibrium climatic states in the global circulation models, so that realistic statistics can be obtained without integration over long time periods.

Exploration of Possible Future Climates

One of the most important applications of climate models is the systematic conduct and evaluation of climatic experiments designed to explore the effects of either natural or anthropogenic changes in the system. It is from such model-based experiments, calibrated with respect to observed behavior, that we must draw our conclusions as to how the climatic system operates and on which we should base our projections of likely future climates. The program in this area should include the determination of the global climatic effects of the following (with both coupled global circulation models and parameterized models):

The *changes of incoming solar radiation*. These experiments should be performed with coupled models, in view of the dominance of the oceans in the planetary heat storage, and should include changes in both the amount and spectral distribution of solar radiation.

The *changes of land surface character and albedo*, as introduced by deforestation, urbanization, irrigation, and changes of agricultural practices.

The *changes of cloudiness*. These experiments should consider the effects of the introduction or removal of both condensation and freezing nuclei and the production of artificial clouds by aircraft.

The *changes of evaporation*, as introduced by reservoirs, irrigation, and transpiration.

The *disposal of waste heat*. These experiments should be made with coupled models and should include a broad range of rates and locations of heat release in both atmosphere and ocean.

The *introduction of dust and particulates* into the troposphere, the stratosphere, or both. These experiments should consider the effects of scattering, absorption, fallout, and scavenging by precipitation and should be designed to simulate the effects of both man-made pollution and volcanic dust.

The partial or complete *removal of the Arctic sea ice* or the Antarctic ice sheet. These experiments should be performed with a coupled model that includes the mass and heat budget of pack ice.

The *diversion of ocean currents*. These experiments should be performed with coupled models.

In climatic simulations of this kind the *physical basis* of each experiment should be carefully examined in order to ensure the adequacy of the particular model or models to be employed. The experiments suggested above are those that we believe should be performed as part of the climatic research program, as they involve processes or areas of likely maximum climatic sensitivity or changes to which the climate's response is relatively uncertain, and/or they represent conceivable (or in some cases likely) future alterations by nature or by man.

It is important in such climatic experiments that the synoptic and statistical significance of the results be carefully examined. This should include the repetition of the experiment under slightly different (but admissible) conditions to determine its stability and noise level and the analysis of independent simulations with other models. Only in this way can we hope to accumulate the necessary experimental knowledge on which to base our expectations of future climatic states. This, together with the knowledge gained from the observational and research portions of the program outlined above, will lay the scientific foundation for what might be called climatic engineering.

Development of Long-Range or Climatic Forecasting

A third important area of application of climatic studies is the problem of long-range or climatic forecasting on time scales of months, seasons, and years. There have been numerous studies of this question almost since the beginning of recorded observations. This research has not solved the problem but has at least identified some of its ingredients. We believe that further efforts should be made to systematically acquire the data and perform the research necessary to attack this problem anew, especially with the aid of climatic models.

Clearly the demand for climatic or long-range forecasts greatly exceeds present capability. An accurate prediction of the temperature or rainfall anomaly over, say, the central plains of North America or over the Ukraine a decade, a year, or even a season in advance would be of great value. And even a somewhat less accurate (but reliable) prediction of the likelihood of such anomalies would be of great use to those involved in agriculture, energy supply allocation, and commerce. At present, the skill of the experimental long-range outlooks prepared by the National Weather Service for the 30-day temperature anomaly at some 100 U.S. cities is only 11 percent greater than chance and only 2 percent greater than chance for the 30-day precipitation anomaly. These forecasts are principally prepared by a mixture of empirical and statistical methods and have also been applied to the seasonal prediction of temperature (Namias, 1968).

The ability of numerical models to perform useful long-range or climatic forecasting (i.e., forecasts over monthly, seasonal, or annual periods) has not been systematically examined because of the large amounts of computation involved and the unavailability of suitable models. Such efforts must also contend with the crucial questions of climatic predictability, noted in Chapter 3, and the long-range stability of the models themselves. We believe that further attention should be given to these problems, using the expanded data base, the coupled dynamical models, and the new computer resources called for in the climatic program. We therefore recommend that

The coupled global circulation models should be systematically applied to the preparation of a series of *long-range forecasts* using observed initial conditions wherever possible. These integrations should extend over at least several seasons, well beyond the limit of local predictability. Appropriate climatic statistics should be drawn from these integrations and systematically compared with the observed variations of all the climatic elements available and statistically analyzed for possibly significant trends of regional climatic anomalies.

The statistical–dynamical models and other appropriate members of the parameterized climate model hierarchy should be used in the preparation of similar long-range forecasts.

Systematic *empirical and diagnostic studies* of longer-period variations in the climatic system should be undertaken with the aid of models and the expanding data base of monitored variables.

Assessment of Climate's Impact on Man

While the above efforts are concerned with the physical aspects of the problem of climatic variation, a climatic research program should also include studies of the impact of climate and climatic change on man himself; this is best done with the guidance and insight provided by climate models. While many studies have been made in this important area, such as those of the Department of Transportation's Climatic Impact Assessment Program (CIAP), more comprehensive research should be undertaken on a long-term basis. These studies may be characterized as seeking answers to such questions as "What is a 1-degree change of mean winter temperature worth, after all?" or even "Climatic variation: so what?" The study of the impacts of climatic variations on man is also a way of establishing priorities for research.

Climate and Food, Water, and Energy

That climate has a dominant influence on agricultural food production, water supply, and the generation and use of energy is generally recognized. The kinds and amounts of crops that may be grown in various regions, the water available for domestic, agricultural, and industrial use, and the consumption of electrical energy and fossil fuels all depend in large measure on the distribution of temperature, rainfall, and sunshine. During the global warming of the first part of this century, for example, the average length of the growing season in England (as measured by the duration of temperatures above 42°F) increased by two to three weeks and during the more recent cooling trend since the 1940's has undergone a comparable shortening (Davis, 1972). Although Maunder (1970), Johnson and Smith (1965), and others have surveyed the vast literature on the effects of climatic change on man, further quantification of these effects is needed, particularly as a function of the time and space scales of atmospheric variability. Accordingly, we recommend that research be devoted to the following:

The systematic assembly from both national and international sources of data on worldwide *food production* and the analysis of their response and sensitivity to variations of climate on monthly and seasonal time scales. Such analyses should then be used to model or simulate the total agricultural response to hypothetical climatic variations. We note that in some cases it may be the variance or extremes of climate, rather than the averages themselves, that will prove to be the more important factor. An applied systems study of this problem has been recently initiated by R. A. Bryson and colleagues at the University of Wisconsin, with the aim of developing predictive relationships between climate and food supply, which will be useful for policy decisions.

The systematic assembly of worldwide data on available *water supply*, both from rainfall and snowpack, and its patterns of use and loss. Analysis should then be undertaken of the water supply system's response and sensitivity to variations of climate and simulation models constructed.

The systematic assembly of worldwide data on the production and use of *energy* and the determination of its response and sensitivity to climatic variations. As in the cases of food and water, simulation models should be constructed, so that the consequences of various patterns of hypothetical climatic change can be estimated.

Social and Economic Impacts

Although it is difficult to obtain useful measures of the social and economic impacts of climatic change, increased attention should be given to this aspect of the problem. This is a problem in which the "noise level" of nonclimatic factors is very high and for which the physical scientist's knowledge must be supplemented by the skills and methods of social and political scientists. The goal of this research should be the development of an overall model of societal response to climatic change. This is an area in which international cooperation should be sought, and efforts such as those now being proposed by the International

Federation of Institutes of Advanced Study should be supported and expanded.

THE PLAN

Our recommendations for the planning and execution of the climatic research program outlined above are given here in terms of what we believe to be the appropriate subprograms, the necessary facilities and support, and the desirable timetable for both the short-range and long-range phases. We also offer some observations on the program's administration and coordination, although we recognize that a program of this scope will require much further planning and that the support and cooperation of many persons and agencies will be necessary for its successful execution.

Subprogram Identification

In a program as broad as that envisaged here, it is convenient to think in terms of a number of components or subprograms, each concerned with a specific portion of the overall effort. Such subprograms also represent the necessary division of effort for the practical execution of the program. The NCRP itself should ensure the coordination of the various subprograms and maintain an appropriate balance of effort among them.

Climatic Data-Analysis Program (CDAP)

In order to promote the extensive assembly and analysis of climatic data outlined above, we recommend that a Climatic Data-Analysis Program (CDAP) be established as a subprogram of the NCRP. The purposes of this subprogram are to facilitate the exchange of data and information among the various climatic data depositories and research projects and to support the coordinated preparation, analysis, and dissemination of appropriate climatic statistics.

Climatic Index Monitoring Program (CIMP)

In order to promote the monitoring of the various climatic indices outlined above, we recommend that a Climatic Index Monitoring Program (CIMP) be established as a second subprogram of the NCRP. The purposes of this subprogram are to support and coordinate the collection of

A NATIONAL CLIMATIC RESEARCH PROGRAM 95

data on selected climatic indices and to ensure their systematic dissemination on a timely and sustained basis.

Climatic Modeling and Applications Program (CMAP)

In order to promote the construction and application of the climatic models outlined above, we recommend that a Climatic Modeling and Applications Program (CMAP) be established as a third subprogram of the NCRP. The purposes of this subprogram are to support and coordinate the development of a broad range of climatic models, to support necessary background scientific research, and to ensure the systematic application of appropriate models to the problems of climatic reconstruction, climatic prediction, and climatic impacts.

Facilities and Support

The availability of adequate facilities and support and the design of coordinating mechanisms are necessary to carry out the various subprograms recommended for the NCRP and should be given careful consideration. Of primary importance are the roles of climatic data-analysis facilities and research consortia, the needed high-speed computers, and the required levels of funding.

Climatic Data-Analysis Facilities

To assist in the implementation of both the Climatic Data Analysis Program (CDAP) and Climatic Index Monitoring Program (CIMP), we recommend the development of new climatic data-analysis facilities at appropriate locations, including linkage to the various specialized data centers and climatic monitoring agencies by a high-speed data-transmission network. Such facilities should have access to machines of the highest speed and capacity available and be staffed by specialists in data analysis, transmission, and display. Collection of certain climatic data by a group of specialized facilities appears more desirable than does collection of all data by a single centralized facility.

We envisage these facilities as performing the bulk of the recommended CDAP. This would include the inventory, compilation, processing, analysis, and documentation of both conventional and proxy climatic data. Close working cooperation is envisaged with specialized data depositories; for conventional atmospheric and oceanic data these include NOAA's National Climatic Center and National Oceanographic

Data Center, for satellite data the National Environmental Satellite Service, for glaciological data the Geological Survey's Data Center A in Tacoma, for ice-core data the Army's Cold Regions Research and Engineering Laboratory, for marine cores Columbia University's Lamont-Doherty Geological Observatory, and for pollen and tree-ring data the universities of Wisconsin and Arizona.

We also envisage the data-analysis facilities as playing a prominent role in the CIMP and in the processing, analysis, and dissemination of the results on as nearly a real-time basis as possible. Certain of the facilities could serve as global climatic "watchdogs" and might have a resident scientific staff to perform diagnostic research as appropriate.

Climatic Research Consortia and Manpower Needs

We envisage the broad range of research and analysis recommended here as being best performed by a number of institutions and groups. This is desirable in order to ensure the breadth of viewpoint and diversity of approach necessary in a problem as close to the unknown as is climatic variation. An attempt to carry out all the recommended activities and research by a single institution would in any case be a practical impossibility.

Research on climate and climatic variation at the present time is principally performed in governmental laboratories and in a variety of research projects in universities and other institutions, usually with the support of the federal government. Chief among the laboratories concerned with elements of the climatic problem are NOAA's Geophysical Fluid Dynamics Laboratory, NOAA's National Environmental Satellite Service and Environmental Data Service, NSF's National Center for Atmospheric Research, and NASA's Goddard Institute for Space Studies. More specialized research on problems related to climate is also performed by the U.S. Geological Survey and by the operational services and laboratories of the U.S. Army, Navy, and Air Force. Many of the climate-related research projects in universities and other institutions are supported by the National Science Foundation through its programs for atmospheric, oceanic, and polar research; by DOT's Climatic Impact Assessment Program; and by ARPA's Climate Dynamics Program. These include the various Quaternary research groups, geological and oceanographic laboratories, numerical modeling groups, and polar studies and environmental institutes.

Each of these efforts makes a contribution to the national climatic research picture, and they represent a valuable reservoir of experience and talent. In order to promote greater cooperation and exchange, to ensure an appropriate balance of effort, and to give such research the needed stability and coherence, we recommend that efforts be made to coordinate present research more effectively as parts of a national climatic research program. We believe that this can be achieved best by the formation of cooperative associations of existing climatic research groups and the initiation of whatever new research efforts may be required as parts of such associations. We accordingly recommend the formation of a number of climatic research consortia among various research groups as appropriate to their interests, with each such consortium having links to computing facilities of the highest speed and capacity available. Such research consortia would serve as valuable coordinating mechanisms for the broad range of climatic research envisaged in the Climatic Modeling and Applications Program (CMAP), as well as giving both coherence and flexibility to the NCRP as a whole. The present MODE, NORPAX, and CLIMAP programs may serve as useful examples for such consortia. As the national program develops, the possible need for new institutional structures or facilities should become clear. Our recommendations reflect the consensus that maximum use should be made of existing institutions while further consideration is given to the possible need for their expansion.

Aside from institutional arrangements, however, we believe that the proposed research program unquestionably calls for the initiation and support of new mechanisms to provide an expanded base of appropriately trained scientific and technical manpower. We accordingly recommend that programs for technical training be developed and that both predoctoral and postdoctoral fellowships in the broad area of climatic research be established as soon as possible.

Computer Requirements

The required access to high-speed computers has been alluded to several times in the discussion of the recommended program. Although it is difficult to make precise projections, the volume of data processing involved in the analysis and monitoring portions of the program alone indicate that a dedicated machine of at least the CDC 7600 class is required for the implementation of the CDAP and CIMP. The computer needs of the research consortia and of the other research groups involved in the modeling portions of the program are even more demanding, in view of the variety of the needed climatic models and tests and the number and the length of the necessary climatic simulation experiments and applications. Our estimates of the NCRP's overall computer requirements are given in Table 6.2 and call for a very significant increase over present levels of computer usage.

If anything, these estimates may be too low. In its computer planning, NCAR has estimated a climate-related usage of several CDC 7600 units by 1980 for the needs of NCAR and the university community it serves (W. M. Washington, personal communication), while the installation of the TI-ASC system at GFDL in 1974 will likely significantly raise their machine usage for climatic studies. As shown in Table 6.2, it is estimated that climatic data analysis and monitoring will require the full-time use of at least one fourth-generation machine, and that climatic modeling and applications will require the full-time use of at least one fifth-generation machine. We therefore recommend that machines of the CDC-7600 class be secured as soon as possible for the use of the data-analysis facilities and the associated elements of the CDAP and CIMP and that planning begin for the acquisition of computers of the TI-ASC or ILLIAC-4 class for the use of the climatic research consortia and the associated elements of the CMAP. It will also be necessary to provide broadband communication links among the various facilities and cooperating groups and with the climatic research community as a whole.

TABLE 6.2 Estimated Computing Needs for the National Climatic Research Program [a]

	Present Use [b]	Projected Use [c]
Climatic data analysis and monitoring	0.2	1.5 [d]
Atmospheric GCM development	0.8 [e]	3.0
Oceanic GCM development	0.2	2.0
Coupled GCM's (climate models)		
Development and tests	0.1	3.0 [f]
Climatic reconstructions	~0	2.0 [f]
Climatic experiments and projections	0.1	3.0 [f]
Other models and studies	0.1	2.0
TOTAL	1.5	16.5

[a] In units of CDC 7600 years.
[b] Estimated 1973 national total, exclusive of operational agencies.
[c] For the program year *circa* 1980.
[d] Envisaged as use by climatic data-analysis facilities.
[e] Estimating 0.2 usage at NCAR, 0.5 usage at GFDL, and 0.1 total usage elsewhere.
[f] Envisaged as use by cooperative climatic research consortia.

Estimated Costs

The cost of the recommended national climatic research program is difficult to determine accurately without a great deal of information on observational, computing, and support costs from the various agencies and institutions presently engaged in the many aspects of climatic research. Rather than seeking such detailed data, we have restricted ourselves to gross projections on the basis of estimates of the costs of present efforts. Our estimates of the expenditures for climatic research (*not* including the costs of instruments, observing platforms, or operational and service-related activities) are given in Table 6.3. Our projections of the growth of these (direct) costs during the early phases of the program (i.e., to the year 1980) are shown in Figure 6.1, along with the percentage increases over the preceding year; these estimates, of course, depend directly on the base figures that are used and are subject to further refinement. These figures are intended for order-of-magnitude guidance only and will require revision as the program develops.

We recognize that the ultimate distribution of resources among the various subprograms of the NCRP will be determined by the sense of priorities of the government and by the capabilities of the research community. The estimates shown in Figure 6.1 for the year 1980 are based on our preception of the needed increases over present efforts in the areas of data analysis and monitoring (CDAP and CIMP), especially those concerning satellite data and the monitoring of oceanic climatic indices. In the area of climatic modeling and applications (CMAP), the largest increases over present efforts are envisaged for the develop-

TABLE 6.3 Estimated Expenditures for Climatic Research[a] (in 10^6/yr)

	Present (1974)	Projected (c.1980)
Climatic data assembly and analysis	5	18
Climatic index monitoring [b]	4	12
Climatic modeling and applications	9	37
	18	67

[a] Based on estimates of the climate-related research sponsored by the NSF, DOT, and DOD and that conducted by GFDL, NCAR, NASA, and NOAA but *not* including essentially operational or service-related activities.
[b] *Not* including costs of instruments or observing platforms.

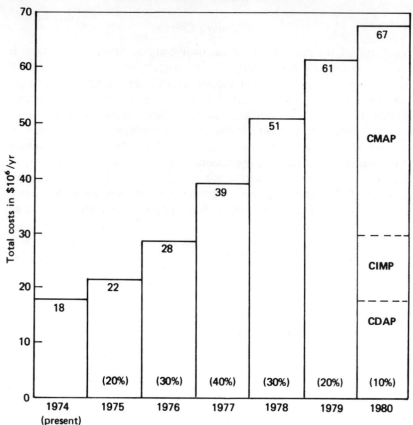

FIGURE 6.1 Projected costs of the National Climatic Research Program (NCRP). The numbers in parentheses are the percent increase over the preceding year's expenditures.

ment and application of coupled global climate models and climatic impact studies. The relatively rapid growth rate during the program's third and fourth years are projected to include the acquisition of the necessary computers and networks. Overall, the recommended program calls for an approximate fourfold expansion of the support of research on climatic variation by the year 1980; the program's costs beyond this time are more difficult to estimate and will depend on the progress and opportunities developed prior to that time.

It is useful to compare these cost projections with the direct and indirect costs of present GARP efforts and those of closely related programs. In fiscal year 1973 the direct GARP expenditures totaled $13.2 million, about 54 percent of which represented expenditures by the Department of Commerce and NASA directed toward the improvement of weather forecasting, with the remainder expended by NSF for research on both forecasting and general circulation studies. Some of these costs are included in the estimates in Table 6.3, insofar as they can be identified as directed toward *climatic* research. The indirect costs associated with GARP amounted to $29.0 million in fiscal year 1973 and are not reflected in the present climatic research estimates.

In addition to these efforts, there are other current programs that contribute to GARP and whose costs should not be overlooked. The implementation of the World Weather Watch (WWW) and its satellite system represented $1.5 million direct costs and $54.5 million indirect costs in fiscal year 1973, while systems design and technological development represented $2.4 million direct costs and $50.1 million indirect costs in the same period. The extent to which elements of the recommended CDAP and CIMP subprograms of the NCRP may be considered as add-ons to such existing programs needs further consideration, as does the extent to which the future costs of GARP itself may be merged with those envisaged for the NCRP.

Also in need of further study are the United States' contributions to the costs of the various subprograms recommended as part of the International Climatic Research Program (ICRP) described below, as well as the impacts of inflation. We also note that funds will be required for the training of additional scientific manpower in all aspects of the research program.

Timetable and Priorities within the Program

We recognize the need for flexibility in a research program of this kind, and that future technological and research discoveries may have important impacts on the direction of climatic research. In spite of these unknown factors, however, some consideration of goals and priorities is useful. Here we present our recommendations for the objectives of the initial phase of the program (1974–1980) and the necessary sequence of planning activities for both these goals and those of the long-term phase (1980–2000). Our recommendations for a coordinated international program are considered subsequently.

The Initial Phase (1974–1980)

Once the decision is made to develop a national climatic research program, we recommend that planning begin immediately for the implementation of its component activities and subprograms. Our specific recommendations for both the immediate and subsequent objectives during this phase of the program are shown in Table 6.4 in terms of the data-analysis, index-monitoring, and modeling subprograms identified earlier. Here our sense of relative priorities is given implicitly by the ranking into immediate and subsequent objectives; these time scales refer to the expected times of the achievement of first useful results, with the recognition that initial development must in some cases begin earlier

TABLE 6.4 Goals for the Initial Phase of the NCRP (1974–1980)

Subprogram	Immediate Objectives (1974–1976)	Subsequent Objectives (1976–1980)
Climatic data analysis (CDAP)	1. Development of climatic data-analysis facilities 2. Statistical analysis of climatic variability, predictability, feedback processes 3. Statistical climatic-impact studies (crops, human affairs)	1. Development of global climatic data-analysis system (FGGE) 2. Assembly and processing of global climatic data (conventional, satellite, historical, proxy data) 3. Development of climatic impact models
Climatic index monitoring (CIMP)	1. Monitoring of oceanic mixed-layer 2. Monitoring of ice, snow, and cloud cover 3. Expansion of proxy data sources 4. Monitoring system simulation studies	1. Satellite monitoring of global heat-balance components 2. Monitoring selected physical processes (FGGE) 3. Development of global climatic index monitoring system
Climatic modeling and applications (CMAP)	1. Development of oceanic mixed-layer models 2. Development and analysis of provisionally coupled GCM's (sensitivity, predictability studies) 3. Development of simplified climatic models and related theoretical studies 4. Selected paleoclimatic reconstructions	1. Development of fully coupled atmosphere–ocean–ice GCM's 2. Development of statistical–dynamical climate models 3. Parameterization of mesoscale processes, simulation of climatic feedback mechanisms (FGGE) 4. Experimental seasonal climatic forecasts by dynamical models

and that further development and application will continue later. This ranking also reflects a balance between the relative ease of accomplishment and the relative potential for initial practical usefulness. We believe that progress toward the subsequent objectives will require the support of *all* immediate objectives of the program, with new priorities evolving as a function of achievement and opportunity.

Relationship to the FGGE (1978–1979)

The First GARP Global Experiment (FGGE), now planned for 1978–1979, is primarily an attempt to collect a definitive global data set for use in the improvement of weather prediction by numerical atmospheric models. The potential value of these data for climatic research lies not so much in their display of seasonal and interhemispheric variations, valuable as that will be, but in the fact that many of the short-period physical processes to be intensely measured or parameterized in FGGE are also important for the understanding of climate. Among these are the processes of convection, boundary-layer dynamics, and the atmosphere's interaction with the surface of the ocean.

The observational requirements during the FGGE call for measurement of the atmospheric temperature, water vapor, cloud cover and elevation, wind, and surface pressure, together with the surface boundary variables of sea-surface temperature, soil moisture, precipitation, snow depth, and sea-ice distribution. To enhance their value for climatic studies, we recommend that these data be supplemented during FGGE insofar as possible by observations of the global distributions of ozone, particulates, surface and planetary albedo, incoming solar and outgoing terrestrial radiation, vegetal cover, and the continental freshwater runoff. We recommend that special observations also be made in conjunction with regional programs, such as NORPAX and POLEX, which are expected to be in operation during the FGGE.

The Long-Term Phase (1980–2000)

The long-range goals and full-scale operation of the NCRP in the period beyond 1980 are portrayed in the upper part of Figure 6.2. During this period, the full interaction among the observational, analysis, modeling, and theoretical components of the program will occur, leading to the development of an operational global climatic data system and, it is hoped, to the acquisition of an increasingly accurate theory of climatic variation. Although priorities cannot be set at such long range, the eventual practical payoffs of this program will be the determination of

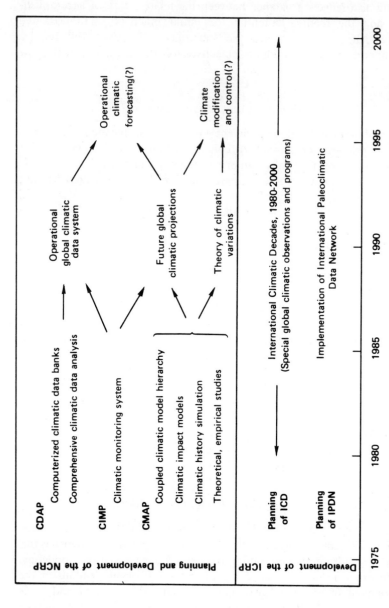

FIGURE 6.2 Long-range goals and activities of the NCRP and coordination with the International Climatic Research Program (ICRP).

the degree to which climatic variations on seasonal, annual, decadal, and longer times scales may be predicted and the degree to which they may be controlled by man.

Administration and Coordination

The administrative structure and coordination of the recommended program are the responsibility of the federal government and were not given extensive consideration. However, noting the concern with the problem of climatic variation in many parts of the government, and the widespread participation of many governmental and nongovernmental groups in climatic research, we believe that the program should be administered in such a way that the interests of all are effectively represented and coordinated. It is particularly important that the advice of the scientific community be used in the design and development of the major elements of the research program.

Both the short-term and long-term goals of the NCRP are also shared by the International Climatic Research Program (ICRP) recommended below. The development of this international program should proceed in parallel with the NCRP and should be closely coordinated with GARP. The principal activities within GARP up to the present time have focused on the problem of improving the accuracy and extending the range of weather forecasts, and the United States' contributions to GARP in particular have emphasized the development and use of numerical models for this purpose. These efforts are necessary steps in the development of an adequate modeling capability for both weather prediction *and* climate, and were they not already under way as part of GARP they would have had to be undertaken through some other means as a prelude to the climatic research program.

A COORDINATED INTERNATIONAL CLIMATIC RESEARCH PROGRAM (ICRP)

Many of the efforts envisaged within the NCRP are of an obvious international character, and the degree to which these should be regarded as national as opposed to international activities is not of critical importance for our purposes. The important point is that there *are* international efforts now under way within GARP of direct relevance to the climatic problem, of which we note especially the International Study Conference on the Physical Basis of Climate and Climate Modeling held in Sweden in July and August 1974 under the auspices of the ICSU/WMO GARP Joint Organizing Committee. The recommendations and programs resulting from this and subsequent planning conferences should be closely coordinated with the U.S. national program. We offer here our recommendations for an appropriate international climatic research program and some observations on how such a program might best be coordinated with GARP itself.

Program Motivation and Structure

The observational programs planned in support of GARP offer an unparalleled opportunity to observe the global atmosphere, and every effort should be made to use these data for climatic purposes as well as for the purposes of weather prediction. The climatic system, however, consists of important nonatmospheric components, including the world's oceans, ice masses, and land surfaces, together with elements of the biosphere. While it is not necessary to measure all of these components in the same detail with which we observe the atmosphere, their roles in climatic variation must not be overlooked.

In addition to the fundamental physical differences discussed in Chapter 3, the problem of climatic variation also differs from that of weather forecasting by the nature of the data sets required. The primary data needs of weather prediction are accurate and dense synoptic observations of the atmosphere's present (and future) states, while the data needed for studies of climatic variation are longer-term statistics of a much wider variety of variables. When climatic variations over long time scales are considered, these variables must be supplied from fields outside of observational meteorology. Thus, an essential characteristic of climate studies is its involvement of a wide range of nonatmospheric scientific disciplines.

The types of numerical models needed for climatic research also differ from those of weather prediction. The atmospheric GCM's (which represent the ultimate in weather models) do not need a time-dependent ocean for weather-forecasting purposes over periods of a week or two. For climatic change purposes, on the other hand, such numerical models *must* include the changes of the oceanic heat storage. Such a slowly varying feature may be regarded as a boundary or external condition for weather prediction but becomes an internal part of the system for climatic variation.

International Climatic Research and GARP

In view of these characteristics, we suggest that while the GARP concern with climate is a natural one, as indicated above the problem of climate

A NATIONAL CLIMATIC RESEARCH PROGRAM 107

goes much beyond the present basis and emphasis of GARP. Accordingly, we recommend that the global climate studies that are under way within GARP be viewed as leading to the organization of a new and long-term international program devoted specifically to the study of climate and climatic variation, which we suggest be called the International Climatic Research Program (ICRP).

International Climatic Decades (1980–2000)

We suggest that the observational programs of GARP, and especially those of the FGGE, be viewed as preliminary efforts, later to be expanded and maintained on a long-term basis. In particular, we recommend that the special data needs of climatic studies be supported on an international scale through the designation of the period 1980–2000 as the International Climatic Decades (ICD), during which intensive efforts would be made to secure as complete a global climatic data base as possible.

The general outline of the envisaged international program (ICRP) is sketched in the lower part of Figure 6.2, and the program's scientific elements are discussed in more detail below.

Program Elements

Climatic Data Analysis

The main thrust of the international climatic program should be the collection and analysis of climatic data during the ICD's, 1980–2000. During this period, the participation of all nations should be sought in order to develop global climatic statistics for a broad set of climatic variables. We urge that these efforts include international cooperation in the systematic summary of all available meteorological observations of climatic value, including oceanographic observations in the waters of coastal nations.

International Paleoclimatic Data Network (IPDN)

We urge the development of an international cooperative program for the monitoring of selected climatic indices and the extraction of historical and proxy climatic data unique to each nation, such as indices of glaciers, rain forest precipitation, lake levels, local desert history, tree rings, and soil records. Specifically, we recommend that this take the form of an International Paleoclimatic Data Network (IPDN), as a subprogram of the ICRP. The cooperation of such organizations as SCAR, SCOR, and the International Union for Quaternary Research (INQUA) should be sought in this program.

The contents of these international observational efforts might possibly broadly follow those recommended for the U.S. national effort, with modifications as appropriate to each nation's needs and capabilities. In addition, we recommend that the ICRP undertake the following:

The international collection of special climatic data sets on such events as widespread drought and floods and following major environmental disturbances such as volcanic eruptions;

Programs to encourage international exchange of climatic data and analyses.

Climatic Research

Although cooperative research studies are desirable, we recognize that the large-scale numerical simulation of climate with CGCM's can now be carried out in only a relatively few countries. To promote wider international participation in climatic research, we therefore recommend that the ICRP include the following:

Programs and activities to encourage international cooperation in climatic research and to facilitate the participation of developing nations that do not yet have adequate training or research facilities.

Internationally supported regional climatic studies in order to describe and model local climatic anomalies of special interest.

The contents of these and other research activities of the ICRP might also broadly follow those recommended for the U.S. national effort, with appropriate modifications for each nation's interests and capabilities.

Global Climatic Impacts

While all nations are tied in some fashion to the world pattern of climate, some are more vulnerable to climatic variations than others by virtue of their locations and the delicacy of their climatic balance. We therefore recommend that the ICRP include the following:

International cooperative programs to assess the impacts of observed climatic changes on the economies of the world's nations, including the effects on the water supply, food production, and energy utilization. This should include the impacts of variations of oceanic climate for those nations whose economies are dependent on the sea. The cooperation of appropriate international agencies of the United Nations and of other groups such as the International Federation of Institutes of Advanced Study should be sought.

Cooperative analyses of the regional impacts of possible future climates. Such studies could be of great importance to many countries, particularly emerging nations making long-range policy decisions concerning the development of their resources.

Program Support

The question of the details of support of the ICRP was not dealt with. It seems clear, however, that an appropriate balance of effort should be maintained among ICRP, the various national climatic research programs, and other international programs such as the World Weather Watch (WWW) and the United Nations Environment Program (UNEP). The services of groups performing the function of the present GARP Joint Organizing Committee and its Joint Planning Staff will also be necessary for the success of the international program.

In order to assist in the coordination of the ICRP, we urge that support be made available by the appropriate agencies of the United Nations on a scale commensurate with the breadth and importance of the problem. This should include a budget adequate for the effective international coordination of the ICRP on a scale significantly greater than that of GARP and on a continuing long-term basis. We also urge that scientific assistance be sought from the International Council of Scientific Unions in support of selected ICRP subprograms.

REFERENCES

Adem, J., 1970: Incorporation of advection of heat by mean winds and by ocean currents in a thermodynamic model for long-range weather prediction, *Mon. Wea. Rev.*, 98:776–786.
Alexander, R. C. 1974: Ocean circulation and temperature prediction model, The Rand Corporation, Santa Monica, Calif. (in preparation).
Alexander, R. C., and R. L. Mobley, 1974: Updated global monthly mean ocean surface temperatures, R–1310–ARPA, The Rand Corporation, Santa Monica, Calif. (in preparation).
Alyea, F. N., 1972: Numerical simulation of an ice age paleoclimate, *Atmospheric Science Paper No. 193*, Dept. of Atmospheric Sciences, Colorado State U., Fort Collins, Colo., 120 pp.
AMTEX Study Group, 1973: The air-mass transformation experiment, *GARP Publ. Series, No. 13*, World Meteorological Organization, Geneva, 55 pp.
Andrews, J. T., R. G. Barry, R. S. Bradley, G. H. Miller, and L. D. Williams, 1972: Past and present glaciological responses to climate in Eastern Baffin Island, *Quaternary Res.*, 2:303–314.
Angell, J. K., J. Korshover, and G. F. Cotten, 1969: Quasibiennial variations in the "centers of action," *Mon. Wea. Rev.*, 97:867–872.
Atwater, M. A., 1972: Thermal effects of urbanization and industrialization in the boundary layer: a numerical study, *Boundary-Layer Meteorol.*, 3:229–245.
Baier, W., 1973: Crop-weather analysis model: review and model development, *J. Appl. Meteorol.*, 12:937–947.
Bathen, K. H., 1972: On the seasonal changes in the depth of the mixed layer in the North Pacific ocean, *J. Geophys. Res.*, 77:7138–7150.
Berggren, W. A., 1971: Tertiary boundaries and correlations, in *The Micropaleontology of Oceans*, Cambridge U. P., London, pp. 693–809.
Berggren, W. A., 1972: A Cenozoic time-scale—some implications for regional geology and paleobiogeography, *Lethaia*, 5:195–215.

Bernabo, J. C., T. Webb, III, and J. McAndrews, 1974: Postglacial isopollen maps of major forest genera and herbs in northeastern North America (in preparation).
Bernstein, R. L., 1974: Mesoscale ocean eddies in the North Pacific: westward propagation, *Science, 183*:71–72.
Bjerknes, J., 1969: Atmospheric teleconnections from the equatorial Pacific, *Mon. Wea. Rev., 97*:163–172.
Black, R. F., R. P. Goldthwait, and H. B. Willman, eds., 1973: The Wisconsinan Stage, *Geol. Soc. Am. Mem. 136,* 334 pp.
Bloom, A. L., 1971: Glacial-eustatic and isostatic controls of sea level since the last glaciation in *The Late Cenozoic Glacial Ages,* K. Turekian, ed., Yale U. P., New Haven, Conn., pp. 355–380.
Born, R., A. Walker, J. Namias, and W. White, 1973: *Monthly mean sea surface temperature departures over the North Pacific Ocean with corresponding subsurface temperature departures at ocean weather stations "Victor," "Papa" and "November" from 1950 to 1970,* SIO Reference Series, 73–28, Scripps Institution of Oceanography, La Jolla, Calif., 248 pp.
Brier, G. W., 1968: Long-range prediction of the zonal westerlies and some problems in data analysis, *Rev. Geophys., 6*:525–551.
Broecker, W. S., and J. van Donk, 1970: Insolation changes, ice volumes, and the O^{18} record in deep-sea cores, *Rev. Geophys. Space Phys., 8*:169–198.
Broecker, W. S., and A. Kaufman, 1965: Radiocarbon chronology of Lake Lahontan and Lake Bonnerville II, Great Basin, *Bull. Geol. Soc. Am., 76*:537–566.
Bryan, K., 1969: Climate and the ocean circulation. Part III. The ocean model, *Mon. Wea. Rev., 97*:806–827.
Bryan, K., and M. D. Cox, 1968: A nonlinear model of an ocean driven by wind and differential heating. Parts I and II, *J. Atmos. Sci., 25*:945–978.
Bryan, K., and M. D. Cox, 1972: The circulation of the world ocean: a numerical study. Part I, a homogeneous model, *J. Phys. Oceanog., 2*:319–335.
Bryan, K., S. Manabe, and R. C. Paconowski, 1974: Global ocean–atmosphere climate model. Part II. The oceanic circulation, Geophysical Fluid Dynamics Laboratory/NOAA, Princeton U., Princeton, N.J., 55 pp. *J. Phys. Oceanog.* (to be published).
Bryson, R. A., 1974: Perspectives on climatic change (to be published in *Science*).
Bryson, R. A., and P. Julian, eds., 1963: Proceedings of the Conference on the Climate of the 11th and 16th Centuries, *NCAR Tech. Note 63–1,* National Center for Atmospheric Research, Boulder, Colo., 103 pp.
Bryson, R. A., and W. M. Wendland, 1970: Climatic effects of atmospheric pollution, in *Global Effects of Environmental Pollution,* S. F. Singer, ed., Springer-Verlag, New York, pp. 13–138.
Budyko, M. I., 1956: *Heat Balance of the Earth's Surface,* U.S. Weather Bureau, Washington, D.C., 259 pp.
Budyko, M. I., 1963: *The Heat Budget of the Earth,* Hydrometeorological Publishing House, Leningrad, 69 pp.
Budyko, M. I., 1969: The effect of solar radiation variations on the climate of the earth, *Tellus, 21*:611–619.
Budyko, M. I., 1971: *Climate and Life,* Hydrometeorological Publishing House, Leningrad.
Budyko, M. I., 1972a: The future climate, *EOS (Trans. Am. Geophys. Union), 53*: 868–874.

Budyko, M. I., 1972b: Comments on "Intransitive model of the earth–atmosphere–ocean system," *J. Appl. Meteorol., 11*:1150.

Butzer, K. W., 1971: *Environment and Archeology*, Aldine-Atherton, Chicago, Ill., 703 pp.

Butzer, K. W., G. L. Isaac, J. L. Richardson, and C. Washbourn-Kamau, 1972: Radiocarbon dating of East African lake levels, *Science, 175*:1069–1076.

Carpenter, R., 1965: *Discontinuity in Greek Civilization*, Cambridge U. P., London.

Chahine, M. T., 1974: Remote sounding of cloudy atmospheres. I. The single cloud layer, *J. Atmos. Sci., 31*:233–243.

Chervin, R., W. L. Gates, and S. H. Schneider, 1974: The effect of time averaging on the noise level of climatological statistics generated by atmospheric general circulation models, 9 pp. (unpublished).

Chylék, P., and J. A. Coakley, Jr., 1974: Aerosols and climate, *Science, 183*:75–77.

Clapp, P. F., 1970: Parameterization of macroscale transient heat transport for use in a mean-motion model of the general circulation, *J. Appl. Meteorol., 9*:554–563.

Committee on Polar Research, 1970: Polar glaciology, in *Polar Research: A Survey*, National Academy of Sciences, Washington, D.C., pp. 73–102.

Corby, G. A., A. Gilchrist, and R. L. Newson, 1972: A general circulation model of the atmosphere suitable for long period integrations, *Q. J. R. Meteorol. Soc., 98*:809–832.

COSPAR Working Group 6, 1972: Report on the application of space techniques to some environmental problems. Preliminary observing system considerations for monitoring important climate parameters. Prepared for The Scientific Committee on Problems of the Environment (SCOPE), 87 pp.

Cox, M. D., 1970: A mathematical model of the Indian Ocean, *Deep-Sea Res., 17*:47–75.

Cox, M. D., 1974: A baroclinic numerical model of the world ocean: preliminary results, *Numerical Models of the Ocean Circulation*, Proceedings of Symposium Held at Durham, New Hampshire, October 17–20, 1972, National Academy of Sciences, Washington, D.C. (in press).

Cox, S. K., 1971: Cirrus clouds and the climate, *J. Atmos, Sci., 28*:1513–1515.

Crowley, W. P., 1968: A global numerical ocean model, *J. Comput. Phys., 3*:111–147.

Crutcher, H. L., and J. M. Meserve, 1970: *Selected level heights, temperatures and dew points for the Northern Hemisphere*, NAVAIR 50–1C–52, Naval Weather Service, Washington, D.C., 370 pp.

Currey, J., 1965: Late Quaternary history, continental shelves of the United States, in *The Quaternary of the United States*, H. E. Wright and D. G. Frey, eds., Princeton U. P., Princeton, N.J., pp. 723–735.

Dansgaard, W., S. J. Johnsen, J. Möller, and C. C. Langway, 1969: One thousand centuries of climatic record from Camp Century on the Greenland ice sheet, *Science, 166*:377–381.

Dansgaard, W., S. J. Johnsen, H. B. Clausen, and C. C. Langway, Jr., 1971: Climatic record revealed by the Camp Century ice core, in *The Late Cenozoic Glacial Ages*, K. Turekian, ed., Yale U. P., New Haven, Conn., pp. 267–306.

Dansgaard, W., S. J. Johnsen, H. B. Clausen, and N. Gunderstrup, 1973: Stable isotope glaciology, *Meddelelser om Grønland, Komm. Vidensk. Under. i Grønland, 197*, 53 pp.

Davis, M. B., 1969: Climatic changes in southern Connecticut recorded by pollen deposition at Rogers Lake, *Ecology, 50*:409–422.
Davis, N. E., 1972: The variability of the onset of spring in Britain, *Q. J. R. Meteorol. Soc., 98*:763–777.
Deardorff, J. W., 1972: Parameterization of the planetary boundary layer for use in general circulation models, *Mon. Wea. Rev., 100*:98–106.
Delsol, F., K. Miyakoda, and R. H. Clarke, 1971: Parameterized processes in the surface boundary layer of an atmospheric circulation model, *Q. J. R. Meteorol. Soc., 97*:181–208.
Denman, K. L., 1973: A time-dependent model of the upper ocean, *J. Phys. Oceanog., 3*:173–184.
Denman, K. L., and M. Miyake, 1973: Upper layer modification at ocean station Papa: observations and simulation, *J. Phys. Oceanog., 3*:185–196.
Denton, G. H., and W. Karlén, 1973: Holocene climatic changes, their pattern and possible cause, *Quaternary Res., 3*:155–205.
Denton, G. H., R. K. Armstrong, and M. Stuiver, 1971: The late Cenozoic glacial history of Antarctica, in *The Late Cenozoic Glacial Ages*, K. Turekian, ed., Yale U. P., New Haven, Conn., pp. 267–306.
Donn, W. L., and M. Ewing, 1968: The theory of an ice-free Arctic Ocean, *Meteorol. Monogr., 8*:100–105.
Douglas, R. G., and S. M. Savin, 1973: *Initial Reports of the Deep Sea Drilling Project, 17*, U.S. Govt. Printing Office, Washington, D.C., pp. 591–605.
Dreimanis, A., and P. F. Karrow, 1972: Glacial history of the Great Lakes—St. Lawrence Region; the classification of the Wisconsinan stage and its correlatives, *24th Int. Geol. Congress* (Section 12, Quaternary Geology), pp. 5–15.
Dwyer, H. A., and T. Petersen, 1973: Time-dependent global energy modeling, *J. Appl. Meteorol., 12*:36–42.
Emiliani, C., 1955: Pleistocene temperatures, *J. Geol., 63*:538–578.
Emiliani, C., 1968: Paleotemperature analysis of Caribbean cores P6304–8 and P6304–9 and a generalized temperature curve for the past 425,000 years, *J. Geol. 74*:109–126.
Environmental Data Service, 1973: *User's Guide to NODC's Data Services*, Key to Oceanographic Records Documentation No. 1, National Oceanic and Atmospheric Administration, Washington, D.C., 72 pp.
Faegre, A., 1972: An intransitive model of the earth–atmosphere–ocean system, *J. Appl. Meteorol., 11*:4–6.
Farrand, W. R., 1971: Late Quaternary paleoclimates of the eastern Mediterranean area, in *The Late Cenozoic Glacial Ages*, K. Turekian, ed., Yale U. P., New Haven, Conn., pp. 529–564.
Ferguson, C. W., 1970. Bristlecone pine chronology and calibration of the radiocarbon time scale, in *Tree Ring Analysis with Special Reference to Northwest America*, H. G. Smith and J. Worrall, eds., Bull. No. 7, Faculty of Forestry, U. of British Columbia, pp. 88–91.
Fleming, R. J., 1972: Predictability with and without the influence of random external forces, *J. Appl. Meteorol., 11*:1155–1163.
Fletcher, J. O., 1972: Ice on the ocean and world climate, in *Beneficial Modification of the Marine Environment*, National Academy of Sciences, Washington, D.C., pp. 4–49.
Fletcher, J. O., Y. Mintz, A. Arakawa, and T. Fox, 1972: Numerical simulation of the influence of Arctic sea ice on climate, in *Energy Fluxes over Polar Surfaces* (Proc. IAMAP/IAPSO/SCAR/WMO Symposium, Moscow, August 3–5, 1971),

REFERENCES

WMO Tech. Note No. 129, World Meteorological Organization, Geneva, pp. 181–218.
Flint, R. F., 1971: *Glacial and Quaternary Geology*, Wiley, New York, 892 pp.
Fritts, H. C., 1971: Dendroclimatology, and dendro-ecology, *Quaternary Res., 1*: 419–449.
Fritts, H. C., 1972: Tree rings and climate, *Sci. Am., 226*:92–100.
Fritts, H. C., T. J. Blasing, B. P. Hayden, and J. E. Kutzbach, 1971: Multivariate techniques for specifying tree-growth and climate relationships and for reconstructing anomalies in paleoclimate, *J. Appl. Meteorol., 10*:845–864.
Frye, C., and H. B. Willman, 1973: Wisconsinan climatic history interpreted from Lake Michigan lobe deposits and soils, *Geol. Soc. Am. Mem. 136*, pp. 135–152.
Fuglister, F. C., 1960: Atlantic Ocean atlas of temperature and salinity profiles and data from the International Geophysical Year of 1957–1958, Vol. 1, Woods Hole Oceanographic Institution, Woods Hole, Mass., 209 pp.
Funnell, B. M., and W. R. Riedel, eds., 1971: *The Micropaleontology of Oceans*, Cambridge U. P., London, 828 pp.
Galt, J. A., 1973: A numerical investigation of Arctic Ocean dynamics, *J. Phys. Oceanog., 3*:379–396.
Gardner, J. V., and J. D. Hays, 1974: The eastern equatorial Atlantic: sea-surface temperature and circulation response to global climatic change during the past 200,000 years, *Geol. Soc. Am., Spec. Paper* (in press).
GARP, 1972: The role of buoys for observations over the ocean areas for the FGGE, Planning Conference on the First GARP Global Experiment, GARP Document 4, World Meteorological Organization, Geneva.
GARP Joint Organizing Committee, 1972: The importance of GARP and the first GARP global experiments for the study of climatic variability and change, Planning Conference on the First GARP Global Experiment, World Meteorological Organization, Geneva, 5 pp.
GARP Joint Organizing Committee, 1973: The first GARP global experiment: objectives and plans, *GARP Publ. Series No. 11*, World Meteorological Organization, Geneva, 107 pp.
GARP Joint Organizing Committee, 1974: Modelling for the first GARP global experiment, *GARP Publ. Series, No. 14*, World Meteorological Organization, Geneva, 261 pp.
Gates, W. L., 1972: The January global climate simulated by the two-level Mintz-Arakawa model: a comparison with observation, R–1005–ARPA, The Rand Corporation, Santa Monica, Calif., 107 pp.
Gates, W. L., 1974: The climatic noise levels generated by a global general circulation model, The Rand Corporation, Santa Monica, Calif. (in preparation).
Gavrilin, B. L., and A. S. Monin, 1970: Calculation of climatic correlations from numerical modeling of the atmosphere, *Izv. Atmos. Oceanic Phys. 6*:659–665.
Gilman, D. L., F. J. Fuglister, and J. M. Mitchell, Jr., 1963: On the power spectrum of "red noise," *J. Atmos. Sci., 20*:182–184.
Green, J. S. A., 1970: Transfer properties of the large-scale eddies and the general circulation of the atmosphere, *Q. J. R. Meteorol. Soc., 96*:157–185.
Haefele, W., 1974: Energy systems, International Institute for Applied Systems Analysis, Baden, Austria, 39 pp. (to be published in *Am. Scientist*).
Haney, R. L., 1974: A numerical study of the large-scale response of an ocean circulation to surface-heat and momentum flux, *J. Phys. Oceanog.* (to be published).
Hays, J. D., T. Saito, N. D. Opdyke, and L. H. Burckle, 1969: Pliocene–Pleisto-

cene sediments of the equatorial Pacific, their paleomagnetic biostratigraphic and climatic record, *Bull. Geol. Soc. Am., 80*:1481–1514.

Hellerman, S., 1967: An updated estimate of the wind stress on the world ocean, *Mon. Wea. Rev., 95*:607–626 (see also *Mon. Wea. Rev., 96*:63–74).

Heusser, C. J., 1966: Late-Pleistocene pollen diagrams from the province of Llanquihue, Southern Chile, *Proc. Am. Phil. Soc., 110*:269–305.

Heusser, C. J., and L. E. Florer, 1974: Correlation of marine and continental Quaternary pollen records from the northeast Pacific and western Washington, *Quaternary Res.* (in press).

Hobbs, P. V., H. Harrison, and E. Robinson, 1974: Atmospheric effects of pollutants, *Science, 183*:909–915.

Holland, J. Z., 1972: Comparative evaluation of some BOMEX measurements of sea surface evaporation, energy flux and stress, *J. Phys. Oceanog., 2*:476–486.

Holland, W. R., 1973: Baroclinic and topographic influences on the transport in western boundary currents, *Geophys. Fluid Dynam., 4*:187–210.

Holland, W. R., and A. D. Hirschman, 1972: A numerical calculation of the circulation in the North Atlantic ocean, *J. Phys. Oceanog., 2*:336–354.

Holloway, J. L., Jr., and S. Manabe, 1971: Simulation of climate by a global general circulation model. I. Hydrologic cycle and heat balance, *Mon. Wea. Rev., 99*:335–370.

Houghton, D. D., 1972: Spatial variations in atmospheric predictability, *J. Atmos. Sci., 29*:816–826.

Houghton, D. D., 1974: The central programme for the GARP Atlantic Tropical Experiment (GATE), *GATE Report No. 3*, World Meteorological Organization, Geneva, 35 pp.

Houghton, D. D., J. E. Kutzbach, M. McClintock, and D. Suchman, 1973: Response of a general circulation model to a sea temperature perturbation, U. of Wisconsin, Madison, 39 pp.

Huang, J. C. K., 1973: *A multi-layer, nonlinear regional dynamical model of the North Pacific Ocean*, Scripps Institution of Oceanography, La Jolla, Calif., 45 pp.

Hughes, T., 1973: Is the West Antarctic ice sheet disintegrating? *J. Geophys. Res., 78*:7884–7910.

Hunkins, K., A. W. H. Bé, N. D. Opdyke, and G. Mathieu, 1971: The late Cenozoic history of the Arctic Ocean, in *The Late Cenozoic Glacial Ages*, K. Turekian, ed., Yale U. P., New Haven, Conn., pp. 215–238.

Imbrie, J., and N. G. Kipp, 1971: A new micropaleontological method for quantitative paleoclimatology: application to a late Pleistocene Caribbean core, in *The Late Cenozoic Glacial Ages*, K. Turekian, ed., Yale U. P., New Haven, Conn., pp. 71–182.

Imbrie, J., J. van Donk, and N. G. Kipp, 1973: Paleoclimatic investigation of a late Pleistocene Caribbean deep-sea core: comparison of isotopic and faunal methods, *Quaternary Res., 3*:10–38.

International Decade of Ocean Exploration (IDOE), 1973: Progress Report, Vol. 2, National Science Foundation, Washington, D.C., 63 pp.

International Glaciological Programme for the Antarctic Peninsula (IGPAP), 1973: Report of a meeting held in Cambridge, England, April 27–30, 15 pp. (unpublished).

ISOS Planning Committee, 1973: International Southern Ocean Studies (ISOS), a sequence of ocean dynamics and monitoring experiments, 58 pp. (unpublished draft).

REFERENCES

Johnsen, S. J., W. Dansgaard, H. B. Clausen, and C. C. Langway, Jr., 1972: Oxygen isotope profiles through the Antarctic and Greenland ice sheets, *Nature, 235*:429–434 (see also *Nature, 236*:249).

Johnson, C. G., and L. P. Smith, eds., 1965: *Biological Significance of Climatic Changes in Britain*, Academic Press, New York, 222 pp.

Joint U.S. POLEX Panel, 1974: U.S. Contribution to the Polar Experiment (POLEX): Part 1 POLEX-GARP (North), National Academy of Sciences, Washington, D.C., 119 pp.

Joseph, J. H., A. Manes, and D. Ashbell, 1973: Desert aerosols transported by Khamsinic depressions and their climatic effects, *J. Appl. Meteorol., 12*:792–797.

Kasahara, A., and T. Sasamori, 1974: Simulation experiments with a 12-layer stratospheric global circulation model. II. Momentum balance and energetics in the stratosphere, *J. Atmos. Sci.* (in press).

Kasahara, A., and W. M. Washington, 1971: General circulation experiments with a six-layer NCAR model, including orography, cloudiness and surface temperature calculation, *J. Atmos. Sci., 28*:657–701.

Kasahara, A., T. Sasamori, and W. M. Washington, 1973: Simulation experiments with a 12-layer stratospheric global circulation model. I. Dynamical effect of the earth's orography and thermal influence of continentality, *J. Atmos. Sci., 30*: 1229–1251.

Kasser, P., 1973: *Fluctuation of Glaciers 1965–1970: A Contribution to the IHD*. Compiled for Permanent Service on the Fluctuations of Glaciers of the IUGG-ICSU, Int. Assoc. of Hydrolog. Sci. and UNESCO, Paris, 357 pp.

Keeling, C. D., R. Bacastow, and C. A. Ekdahl, 1974: Diminishing role of the oceans in industrial CO_2 uptake during the next century (in preparation).

Kellogg T. B., 1974: Late Quaternary climatic changes in the Norwegian and Greenland seas, in *Proc. 24th Alaskan Sci. Conf.*, U. of Alaska, Fairbanks, Aug. 15–17, 1973 (to be published).

Kennett, J. P., and P. Huddlestun, 1972: Late Pleistocene paleoclimatology, foraminiferal biostratigraphy, and tephrochronology; Western Gulf of Mexico, *Quaternary Res., 2*:38–69.

Kipp, N. G., 1974: A new transfer function for estimating past sea-surface conditions from the sea-bed distribution of planktonic foraminiferal assemblages, *Geol. Soc. Am. Spec. Paper* (in press).

Kondratyev, K. Ya., 1973: The complete atmospheric energetics experiment (CAENEX), *GARP Publ. Series No. 12*, World Meteorological Organization, Geneva, 43 pp.

Kraus, E. B., 1973: Comparison between ice age and present general circulations, *Nature, 245*:129–133.

Kukla, G. J., 1970: Correlations between loesses and deep-sea sediments, *Geol. Fören. Stockholm Förh., 92*:148–180.

Kukla, G. J., and H. J. Kukla, 1974: Increased surface albedo in the Northern Hemisphere, *Science, 183*:709–714.

Kukla, G. J., R. K. Matthews, and J. M. Mitchell, Jr., 1972: The present interglacial: How and when will it end?, *Quaternary Res., 2*:261–269.

Kuo, H. L., and G. Veronis, 1973: The use of oxygen as a test for an abyssal circulation model, *Deep-Sea Res., 20*:871–888.

Kurihara, Y., 1970: A statistical–dynamical model of the general circulation of the atmosphere, *J. Atmos. Sci., 27*:847–870.

Kurihara, Y., 1973: Experiments on the seasonal variation of the general circulation in a statistical dynamical model, *J. Atmos. Sci.*, 30:25–49.

Kutzbach, J. E., 1970: Large-scale features of monthly mean Northern Hemisphere anomaly maps of sea-level pressure, *Mon. Wea. Rev.*, 98:708–716.

Kutzbach, J. E., and R. E. Bryson, 1974: Variance spectrum of Holocene climatic fluctuations in the North Atlantic sector, Dept. of Meteorol., U. of Wisconsin, Madison, 15 pp. (unpublished).

LaMarche, V. C., Jr., 1974: Paleoclimatic inferences from long tree-ring records, *Science*, 183:1043–1048.

LaMarche, V. C., Jr., and T. P. Harlan, 1973: Accuracy of tree-ring dating of bristlecone pine for calibration of the radiocarbon time scale, *J. Geophys. Res.*, 78:8849–8858.

Lamb, H. H., 1968: *The Changing Climate*, Methuen, London, 236 pp.

Lamb, H. H., 1969: Climatic fluctuations, in *World Survey of Climatology, 2, General Climatology*, H. Flohn, ed., Elsevier, New York, pp. 173–249.

Lamb, H. H., 1970: Volcanic dust in the atmosphere with chronology and assessment of its meteorological significance, *Phil. Trans. R. Soc.*, 266:425–533.

Lamb, H.H., ed., 1972: *Climate: Present, Past and Future*, Vol. 1, Methuen, London, 613 pp.

Lamb, H. H., 1973a: The development of monthly and seasonal weather forecasting in the U.K. Meteorological Office, Working Group on Climatic Fluctuations, Commission for Atmospheric Sciences, World Meteorological Organization, 18 pp.

Lamb, H. H., 1973b: Second Annual Report, Climatic Research Unit, U. of East Anglia, Norwich, England, 28 pp.

Lamb, H. H., and R. A. S. Ratcliffe, 1972: On the magnitude of climatic anomalies in the oceans and some related observations of atmospheric circulation behavior, in *Climate: Present, Past and Future, 1*, H. H. Lamb, ed., Methuen, London, 613 pp.

Lamb, H. H., and A. Woodroffe, 1970: Atmospheric circulation during the last ice age, *Quaternary Res.*, 1:29–58.

Landsberg, H. E., 1970: Man-made climatic changes, *Science*, 170:1265–1274.

Landsberg, H. E., 1973: An analysis of the annual rainfall at Dakar (Senegal), 1887–1972, U. of Maryland, College Park (unpublished).

Langway, C. C., Jr., 1970: Stratigraphic analysis of a deep ice core from Greenland, *Geol. Soc. Am. Spec. Paper 125*, 186 pp.

Leith, C. E., 1971: Atmospheric predictability and two-dimensional turbulence, *J. Atmos. Sci.*, 28:145–161.

Leith, C. E., 1973: The standard error of time-average estimates of climatic means, *J. Appl. Meteorol.*, 12:1066–1069.

Leopold, E. B., 1969: Late Cenozoic palynology, in *Aspects of Palynology*, R. H. Tschudy and R. A. Scott, eds., Interscience, New York, pp. 377–438.

LeRoy Ladurie, E., 1971: *Times of Feast, Times of Famine*, Doubleday, Garden City, N.Y., 426 pp.

Livingstone, D. A., 1971: A 22,000-year pollen record from the plateau of Zambia, *Limnol. Oceanog.*, 16:349–356.

London, J., and T. Sasamori, 1971: Radiative energy budget of the atmosphere, in *Man's Impact on the Climate*, W. H. Matthews, W. W. Kellogg, and G. D. Robinson, eds., MIT Press, Cambridge, Mass., pp. 141–155.

Lorenz, E. N., 1968: Climatic determinism, in Causes of Climatic Change, *Meteorol. Monogr.*, 8:1–3.

Lorenz, E. N., 1969: The predictability of a flow which possesses many scales of motion, *Tellus*, 21:289–307.

Lorenz, E. N., 1970: Climatic change as a mathematical problem, *J. Appl. Meteorol.* 9:325–329.

Lorenz, E. N., 1973: On the existence of extended range predictability, *J. Appl. Meteorol.*, 12:543–546.

Lovins, A. B., 1974: Thermal limits to world energy-use (to be published in *Bull. Atomic Scientists*).

Ludlum, D. L., 1966: *Early American Winters, 1604–1820*, Am. Meteorol. Soc., Boston, Mass., 283 pp.

Ludlum, D. L., 1968: *Early American Winters, 1821–1870*, Am. Meteorol. Soc., Boston, Mass., 257 pp.

Luz, B., 1973: Stratigraphic and paleoclimatic analysis of late Pleistocene tropical southeast Pacific cores, *Quaternary Res.*, 3:56–72.

Ivovitch, M. I., and S. P. Ovtchinnikov, 1964: *Physical Geographic Atlas of the World*, USSR Academy of Sciences and State Geodetic Commission, Moscow.

MacCracken, M. C., 1968: *Ice Age Theory Analysis by Computer Model Simulation*, PhD Dissertation, U. of California, Davis, 193 pp.

MacCracken, M. C., 1972: Zonal atmospheric model ZAM2, Lawrence Livermore Laboratory, U. of California, Livermore (Preprint UCRL-74254), 64 pp.

MacCracken, M. C., and F. M. Luther, 1973: Climate studies using a zonal atmospheric model, Lawrence Livermore Laboratory, U. of California, Livermore (Preprint UCRL-74887), 40 pp.

Machata, L., 1973: Man's influence on the climate. A status report, Commission for Atmospheric Sciences, World Meteorological Organization, Versailles, 19 pp. (unpublished).

Mahlman, J. D., and S. Manabe, 1972: Numerical simulation of the stratosphere: implications for related climate change problems, in *Proc. Survey Conf., CIAP*, U.S. Department of Transportation, pp. 186–193.

Manabe, S., 1969a. Climate and the ocean circulation: I. The atmospheric circulation and the hydrology of the earth's surface, *Mon. Wea. Rev.*, 97:739–774.

Manabe, S., 1969b: Climate and the ocean circulation: II. The atmospheric circulation and the effect of heat transfer by ocean currents, *Mon. Wea. Rev., 97*: 775–805.

Manabe, S., 1971: Estimates of future change of climate due to the increase of carbon dioxide concentration in the air, in *Man's Impact on the Climate*, W. H. Matthews, W. W. Kellogg, and G. D. Robinson, eds., MIT Press, Cambridge, Mass., pp. 249–264.

Manabe, S., and K. Bryan, 1969: Climate calculations with a combined ocean–atmosphere model, *J. Atmos. Sci.*, 26:786–789.

Manabe, S., and R. W. Wetherald, 1967: The thermal equilibrium of the atmosphere with a given distribution of relative humidity, *J. Atmos. Sci.*, 24:241–259.

Manabe, S., J. L. Holloway, and D. G. Hahn, 1972: Seasonal variation of climate in a time-integration of a mathematical model of the amtosphere, in *Proc. Symp. Phys. and Dyn. Climatol.* (Leningrad, 1971), World Meteorological Organization, Geneva.

Manabe, S., K. Bryan, and M. J. Spelman, 1974a: A global atmosphere–ocean climate model. Part 1. The atmospheric circulation, Geophysical Fluid Dynamics Laboratory/NOAA, Princeton U., Princeton, N.J., 76 pp. (to be published).

Manabe, S., D. G. Hahn, and J. L. Holloway, Jr., 1974b: The seasonal variation of the tropical circulation as simulated by a global model of the atmosphere, *J. Atmos. Sci., 31*:43-83.

Manley, G., 1959: Temperature trends in England, 1680-1959, *Arch. Meteorol. Geophys. Bioklimatol., Ser. B., 9*:413-433.

Matthews, R. K., 1973: Relative elevation of late Pleistocene high sea level stands: Barbados uplift rates and their implications, *Quaternary Res., 3*:147-153.

Maunder, W. J., 1970: *The Value of the Weather*, Methuen, London, 388 pp.

Maykut, G. A., and N. Untersteiner, 1971: Some results from a time-dependent thermodynamic model of sea ice, *J. Geophys. Res., 76*:1550-1575.

McIntyre, A., et al., 1972a: The glacial North Atlantic 17,000 years ago, paleoisotherm and oceanographic maps derived from floral-faunal parameters by CLIMAP, in *Geol. Soc. Am., Program Ann. Mtg.*, abstract, pp. 590-591.

McIntyre, A., W. F. Ruddiman, and R. Jantzen, 1972b: Southward penetrations of the North Atlantic polar front; faunal and floral evidence of large-scale surface-water mass movements over the last 225,000 years, *Deep-Sea Res., 19*:61-77.

McIntyre, A., et al., 1974: The glacial North Atlantic 18,000 years ago: a CLIMAP reconstruction, *Geol. Soc. Am. Spec. Paper* (in press).

Mesolella, K. J., R. K. Matthews, W. S. Broecker, and D. L. Thurber, 1969: The astronomic theory of climatic change: Barbados data, *J. Geol., 77*:250-274.

Mid-ocean Dynamics Experiment-one, Scientific Council, 1973: *Mode-1: An Overview of the Program and Detailed Description of the Field Experiment*, National Science Foundation, Washington, D.C., 38 pp.

Milankovitch, M., 1930: Mathematische Klimalehre und astronomische Theorie der Klimaschwankungen, *Handbuch der Klimatologie*, Bd. I, Teil A, Verlag Borntrager, Berlin, 176 pp.

Milliman, J. D., and K. O. Emery, 1968: Sea levels during the past 35,000 years, *Science, 162*:1121-1123.

Mintz, Y., A. Katayama, and A. Arakawa, 1972: Numerical simulation of the seasonally and inter-annually varying tropospheric circulation, in *Proc. Survey Conf., CIAP*, U.S. Department of Transportation, pp. 194-216.

Mitchell, J. M., Jr., 1966: Stochastic models of air-sea interaction and climatic fluctuation, in *Proc. Symp. Arctic Heat Budget and Atmospheric Circulation*, J. O. Fletcher, ed., RM-5233-NSF, The Rand Corporation, Santa Monica, Calif., pp. 45-74.

Mitchell, J. M., Jr., 1970: A preliminary evaluation of atmospheric pollution as a cause of the global temperature fluctuation of the past century, in *Global Effects of Environmental Pollution*, S. F. Singer, ed., Springer-Verlag, New York, pp. 97-112.

Mitchell, J. M., Jr., 1971a: The effect of atmospheric aerosols on climate with special reference to temperature near the earth's surface, *J. Appl. Meteorol., 10*: 703-714.

Mitchell, J. M., Jr., 1971b: The problem of climatic change and its causes, in *Man's Impact on the Climate*, W. H. Matthews, W. W. Kellogg, and G. D. Robinson, eds., MIT Press, Cambridge, Mass., pp. 133-140.

Mitchell, J. M., Jr., 1973a: A reassessment of atmospheric pollution as a cause of long-term changes of global temperature, in *Global Effects of Environmental Pollution*, 2nd ed., S. F. Singer, ed., Reidel, Dordrecht, Holland.

Mitchell, J. M., Jr., 1973b: The natural breakdown of the present interglacial and its possible intervention by human activities, *Quaternary Res., 2*:436–445.

Mitchell, J. M., Jr., 1974: The global cooling effect of increasing atmospheric aerosols, fact or fiction? in *Proc. IAMAP/WMO Symp. Phys. and Dyn. Climatol.* (Leningrad, USSR, August 16–20, 1971), World Meteorological Organization, Geneva (in press).

Möller, F., 1951: Vierteljahrskarten des Niederschlags für die ganze Erde, *Petermanns Geographische Mitteilungen,* Justus Perthes, Gotha, pp. 1–7.

Moore, T. C., Jr., 1973: Late Pleistocene–Holocene oceanographic changes in the northeastern Pacific, *Quaternary Res., 3*:99–109.

Munk, W. H., and J. D. Woods, 1973: Remote sensing of the ocean, Working Group Report II, in *Boundary-Layer Meteorol., 5*:201–209.

Namias, J., 1968: Long-range weather forecasting—history, current status and outlook, *Bull. Am. Meteorol. Soc., 49*:438–470.

Namias, J., 1969: Seasonal interactions between the North Pacific Ocean and the atmosphere during the 1960's, *Mon. Wea. Rev., 97*:173–192.

Namias, J., 1970: Climatic anomaly over the United States during the 1960's, *Science, 170*:741–743.

Namias, J. 1971a: Space scales of sea-surface temperature patterns and their causes, *Fishery Bull., 70*:611–617.

Namias, J., 1971b: The 1968–69 winter as an outgrowth of sea and air coupling during antecedent seasons, *J. Phys. Oceanog., 1*:65–81.

Namias, J., 1972a: Large-scale and long-term fluctuations in some atmospheric and oceanic variables, in *The Changing Chemistry of the Oceans, Nobel Symposium 20,* O. Dryssen and D. Jagner, eds., Wiley, New York, pp. 27–48.

Namias, J., 1972b: Influence of Northern Hemisphere general circulation on drought in northeast Brazil, *Tellus, 24*:336–342.

Namias, J., 1973: Thermal communication between the sea surface and the lower troposphere, *J. Phys. Oceanog., 3*:373–378.

National Science Board, 1972: *Patterns and Perspectives in Environmental Science,* National Science Foundation, Washington, D. C., 426 pp.

Newell, R. E., 1971: The global circulation of atmospheric pollutants: how do they travel and how might they affect world climates, *Sci. Am., 244*:32–47.

Newel, R. E., 1972: Climatology of the stratosphere from observations, *Proc. Survey Conf., CIAP,* U.S. Department of Transportation, pp. 165–185.

Newell, R. E., D. G. Vincent, T. G. Dopplick, D. Ferruzza, and J. W. Kidson, 1969: The energy balance of the global atmosphere, in *The Global Circulation of the Atmosphere,* G. A. Corby, ed., Royal Meteorol. Soc., London, pp. 42–90.

Newell, R. E., J. W. Kidson, D. G. Vincent, and G. J. Boer, 1972: *The General Circulation of the Tropical Atmosphere and Interactions with Extra Tropical Latitudes,* MIT Press, Cambridge, Mass., 258 pp.

Niiler, P. P., and W. S. Richardson, 1973: Seasonal variability of the Florida current, *J. Mar. Res., 31*:144–167.

Olsson, I. U., ed., 1970: *Radiocarbon Variations and Absolute Chronology, Nobel Symposium 12,* Wiley, New York, 652 pp.

Oort, A. H., 1972: Atmospheric circulation statistics for a 5-year period, Geophysical Fluid Dynamics Laboratory, Princeton, 30 pp. (unpublished).

Oort, A. H., and E. M. Rasmusson, 1971: Atmospheric circulation statistics, *NOAA Prof. Paper 5,* U.S. Govt. Printing Office, Washington, D.C., 323 pp.

Paterson, W. S. B., 1972: Laurentide ice sheet: estimated volumes during late Wisconsin, *Rev. Geophys. Space Phys., 10*:885–917.
Paulson, C. A., E. Leavitt, and R. G. Fleagle, 1972: Air–sea transfer of momentum, heat and water determined from profile measurements during BOMEX, *J. Phys. Oceanog., 2*:487–497.
Petukhov, V. K., and Ye. M. Feygel'son, 1973: A model of long-period heat and moisture exchange in the atmosphere over the ocean, *Izv. Atmos. Oceanic Phys., 9*:352–362.
Pike, A. C., 1972: Response of a tropical atmosphere and ocean model to seasonally variable forcing, *Mon. Wea. Rev., 100*:424–433.
Pisias, N. G., J. P. Dauphin, and C. Sancetta, 1973: Spectral analysis of late Pleistocene–Holocene sediments, *Quaternary Res., 3*:3–9.
Pittock, A. B., 1972: How important are climatic changes? *Weather, 27*:262–271.
Plass, G. N., 1956: The carbon dioxide theory of climatic change, *Tellus, 8*:140–154.
Porter, S. C., 1971: Fluctuations of late Pleistocene alpine glaciers in western North America, in *The Late Cenozoic Glacial Ages*, K. Turekian, ed., Yale U. P., New Haven, Conn., pp. 307–330.
Posey, J. W., and P. F. Clapp, 1964: Global distribution of normal surface albedo, *Geofis. Int., 4*:53–58.
Raschke, E., T. H. Vonder Haar, W. R. Bandeen, and M. Pasternak, 1973: The annual radiation balance of the earth–atmosphere system during 1969–1970 from Nimbus 3 measurements, *J. Atmos. Sci., 30*:341–364.
Rasool, S. I., and S. H. Schneider, 1971: Atmospheric carbon dioxide and aerosols: effects of large increases on global climate, *Science, 173*:138–141.
Ratcliffe, R. A. S., 1973: Recent work on sea surface temperature anomalies related to long-range forecasting, *Weather, 28*:106–117.
Ratcliffe, R. A., and R. Murray, 1970: New lag associations between North Atlantic sea temperature and European pressure applied to long-range weather forecasting, *Q. J. R. Meteorol. Soc., 96*:226–246.
Reid, J. L., and W. D. Nowlin, Jr., 1971: Transport of water through the Drake Passage, *Deep-Sea Res., 18*:51–64.
Richmond, G. M., 1972: Appraisal of the future climate of the Holocene in the Rocky Mountains, *Quaternary Res., 2*:315–322.
Roberts, W. O., 1973: Relationships between solar activity and climate change, preprint of paper given at NASA/Goddard, 30 pp. (unpublished).
Roberts, W. O., and R. H. Olson, 1973: New evidence for effects of variable solar corpuscular emission on the weather, *Rev. Geophys. Space Phys., 11*:731–740.
Robinson, G. D., 1971a: The predictability of a dissipative flow, *Q. J. R. Meteorol. Soc., 97*:300–312.
Robinson, G. D., 1971b: Review of climate models, in *Man's Impact on the Climate*, W. H. Matthews, W. W. Kellogg, and G. D. Robinson, eds., MIT Press, Cambridge, Mass., pp. 205–215.
Robinson, M. K., and R. A. Bauer, 1971: Atlas of monthly mean sea surface and subsurface temperature and depth of the top of the thermocline, North Pacific Ocean, Fleet Numerical Weather Central, Monterey, Calif., 97 pp. (unpublished).
Rooth, C., 1972: A linearized bottom friction law for large-scale oceanic motions, *J. Phys. Oceanog., 2*:509–510.
Rowntree, P. R., 1972: The influence of tropical east Pacific Ocean temperatures on the atmosphere, *Q. J. R. Meteorol. Soc., 98*:290–321.

Ruddiman, W. F., 1971: Pleistocene sedimentation in the equatorial Atlantic, stratigraphy and faunal paleoclimatology, *Bull. Geol. Soc. Am., 82*:283–302.

Ruddiman, W. F., and L. G. Glover, 1972: Vertical mixing of ice-rafted volcanic ash in North Atlantic sediments, *Geol. Soc. Am. Bull., 83*:2817–2836.

Sachs, H. M., 1973: North Pacific radiolarian assemblages and their relationship to oceanographic parameters, *Quaternary Res., 3*:73–88.

Saltzman, B., 1973: Parameterizations of hemispheric heating and temperature variance fields in the lower troposphere, *Pure Appl. Geophys., 105*:890–899.

Saltzman, B., and A. D. Vernekar, 1971: An equilibrium solution for the axially symmetric component of the earth's macroclimate, *J. Geophys. Res., 76*:1498–1524.

Saltzman, B., and A. D. Vernekar, 1972: Global equilibrium solutions for the zonally averaged macroclimate, *J. Geophys. Res., 77*:3936–3945.

Sancetta, C., J. Imbrie, and N. G. Kipp, 1973: Climatic record of the past 130,000 years in North Atlantic deep-sea core V23–82: correlations with the terrestrial record, *Quaternary Res., 3*:110–116.

Sasamori, T., 1970: A numerical study of atmospheric and soil boundary layers, *J. Atmos. Sci., 27*:1122–1137.

Sawyer, J. S., 1966: Possible variations of the general circulation of the atmosphere, in *World Climate from 8000 to 0 B.C.*, Royal Meteorol. Soc., London, pp. 218–229.

Sawyer, J. S., 1971: Possible effects of human activity on world climate, *Weather, 26*:251–262.

Schneider, S. H., 1971: A comment on climate: the influence of aerosols, *J. Appl. Meteorol., 10*:840–841.

Schneider, S. H., 1972: Cloudiness as a global climatic feedback mechanism: the effects on the radiation balance and surface temperature of variations in cloudiness, *J. Atmos. Sci., 29*:1413–1422.

Schneider, S. H., and R. E. Dickinson, 1974: Climate modeling, *Rev. Geophys. Space Phys., 12*:447–493.

Schneider, S. H., and T. Gal-Chen, 1973: Numerical experiments in climate stability, *J. Geophys. Res., 78*:6182–6194.

Schneider, S. H., and W. W. Kellogg, 1973: The chemical basis for climate change, in *Chemistry of the Lower Atmosphere*, S. I. Rasool, ed., Plenum, New York, pp. 203–249.

Schwartzbach, M., 1961: The climatic history of Europe and North America, in *Descriptive Paleoclimatology*, A. E. M. Nairn, ed., Interscience, New York, pp. 255–291.

Sellers, W. D., 1969: A global climatic model based on the energy balance of the earth–atmosphere system, *J. Appl. Meteorol., 8*:392–400.

Sellers, W. D., 1973: A new global climatic model, *J. Atmos. Sci., 12*:241–254.

Shackleton, N. J., and J. P. Kennett, 1974a: Oxygen and carbon isotope record at DSDP Site 284 and their implications for glacial development, *Initial Reports of the Deep Sea Drillng Project, 29* (in press).

Shackleton, N. J., and J. P. Kennett, 1974b: Paleotemperature history of the Cenozoic and the initiation of Antarctic glaciation; oxygen and carbon isotope analyses in DSDP sites to 77, 279, and 281, *Initial Reports of the Deep Sea Drilling Project, 29* (in press).

Shackleton, N. J., and N. D. Opdyke, 1973: Oxygen isotope and paleomagnetic stratigraphy of equatorial Pacific core V28–238: oxygen isotope temperatures and ice volumes on a 10^5 and 10^6 year scale, *Quaternary Res., 3*:39–55.

Shenk, W. E., and V. V. Salomonson, 1972: A multispectral technique to determine sea-surface temperature using Nimbus 2 data, *J. Phys. Oceanog.*, 2:157–167.

Smagorinsky, J., 1974: Global atmospheric modeling and the numerical simulation of climate, in *Weather Modification*, W. N. Hess, ed., Wiley, New York (to be published).

Smith, R. F. T., 1973: A note on the relationship between large-scale energy functions and characteristics of climate, *Q. J. R. Meteorol. Soc.*, 99:693–703.

Smith, W. L., D. H. Staelin, and J. T. Houghton, 1973: Vertical temperature profiles from satellites: results from second generation instruments aboard Nimbus-5, COSPAR Symposium on Approaches to Earth Survey Problems Through the Use of Space Techniques, Konstanz, 25 pp. (unpublished).

Somerville, R. C. J., P. H. Stone, M. Halem, J. E. Hansen, J. S. Hogan, L. M. Druyan, G. Russell, A. A. Lacis, W. J. Quirk, and J. Tenenbaum, 1974: The GISS model of the global atmosphere, *J. Atmos. Sci.*, 31:84–117.

Sorenson, C. J., and J. C. Knox, 1973: Paleosoils and paleoclimates related to late Holocene forest/tundra border migrations, in *Proc. Calgary Inst. Conf. on Prehistory and Paleoecology of the Western Arctic and Subarctic* (to be published).

Spar, J., 1973a: Transequatorial effects of sea-surface temperature anomalies in a global general circulation model, *Mon. Wea. Rev.*, 101:554–563.

Spar, J., 1973b: Some effects of surface anomalies in a global general circulation model, *Mon. Wea. Rev.*, 101:91–100.

Starr, V. P., and A. H. Oort, 1973: Five-year climatic trend for the Northern Hemisphere, *Nature*, 242:310–313.

Steinen, R. P., R. S. Harrison, and R. K. Matthews, 1973: Eustatic low stand of sea level between 105,000 and 125,000 B.P.: evidence from the subsurface of Barbados, W.I., *Geol. Soc. Am. Bull.*, 84:63–70.

Stommel, H., 1973: First steps toward the design of an oceanographic program in the Indian Ocean during the first global GARP experiment (unpublished draft).

Stone, P. H., 1973: The effect of large-scale eddies on climatic change, *J. Atmos. Sci.*, 30:521–529.

Sverdrup, H., M. W. Johnson, and R. H. Fleming, 1942: *The Oceans*, Prentice-Hall, Englewood Cliffs, N.J., 1087 pp.

Swain, A., 1973: A history of fire and vegetation in northeastern Minnesota as recorded in lake sediments, *Quaternary Res.*, 3:383–396.

Takano, K., Y. Mintz, and Y. J. Han, 1973: Numerical simulation of the seasonally varying baroclinic world ocean circulation, Dept. of Meteorol., U. of California, Los Angeles (unpublished).

Taljaard, J. J., H. Van Loon, H. L. Crutcher, and R. L. Jenne, 1969: *Climate of the Upper Air: Southern Hemisphere, 1. Temperatures, dew points and heights at selected pressure levels*, NAVAIR 50–1C–55, Naval Weather Service, Washington, D.C.

Tsukada, M., 1968: Vegetation and climate around 10,000 B.P. in central Japan, *Am. J. Sci.*, 265:562–585.

Turekian, K., ed., 1971: *The Late Cenozoic Glacial Ages*, Yale U. P., New Haven, Conn., 606 pp.

U.S. Navy Hydrographic Office, 1944: *World Atlas of Sea Surface Temperatures*, H. O. Publ. 225, Navy Hydrographic Office, Washington, D.C. (see also subsequent H. O. publications).

Van der Hammen, T., T. A. Wijmstra, and W. H. Zagwijn, 1971: The floral record of the late Cenozoic of Europe, in *The Late Cenozoic Glacial Ages*, K. Turekian, ed., Yale U. P., New Haven, Conn., 319–424.

Vernekar, A. D., 1972: Long-period global variations of incoming solar radiation, *Meteorol. Mongr., 12*, No. 34, 128 pp.
Vonder Haar, T. H., and A. H. Oort, 1973: New estimates of annual poleward energy transport by Northern Hemisphere oceans, *J. Phys. Oceanog., 3*:169–172.
Vonder Haar, T. H., and V. E. Suomi, 1971: Measurements of the earth's radiation budget from satellites during a five-year period. Part I. Extended time and space means, *J. Atmos. Sci., 28*:305–314.
Vulis, I. L., and A. S. Monin, 1971: Spectra of long-term fluctuations of meteorological fields, *Doklady Akad. Nauk SSSR, Geophys. Sec., 197*:4–6.
Wagner, A. J., 1971: Long-period variations in seasonal sea-level pressure over the Northern Hemisphere, *Mon. Wea. Rev., 99*:49–69.
Wagner, A. J., 1973: The influence of average snow depth on monthly mean temperature anomaly, *Mon. Wea. Rev., 101*:624–626.
Wahl, E. W., 1972: Climatological studies of the large-scale circulation in the Northern Hemisphere, *Mon. Wea. Rev., 100*:553–564.
Wahl, E. W., and T. L. Lawson, 1970: The climate of the midnineteenth century United States compared to current normals, *Mon. Wea. Rev., 98*:259–265.
Walcott, R. I., 1972: Past sea levels, eustasy and deformation of the earth, *Quaternary Res., 2*:1–14.
Washburn, A. L., 1973: *Periglacial Processes and Environments,* Arnold Press, London, 320 pp.
Washington, W. M., 1972: Numerical climatic-change experiments: the effect of man's production of thermal energy, *J. Appl. Meteorol., 11*:768–772.
Washington, W. M., and L. G. Thiel, 1970: Digitized global monthly mean ocean surface temperatures, *Tech. Note 54,* National Center for Atmospheric Research, Boulder, Colo.
Webb, T., and R. A. Bryson, 1972: Late- and post-glacial climatic change in the northern midwest, USA: quantitative estimates derived from fossil pollen spectra by multivariate statistical analysis, *Quaternary Res., 2*:70–115.
Wendland, W. M., and R. A. Bryson, 1974: Dating climatic episodes of the Holocene, *Quaternary Res.* (to be published).
West, R. G., 1968: *Pleistocene Geology and Biology,* Wiley, New York, 377 pp.
Wetherald, R. T., and S. Manabe, 1972: Response of the joint ocean–atmosphere model to the seasonal variation of the solar radiation, *Mon. Wea. Rev., 100*: 42–59.
Weyl, P. K., 1968: The role of the oceans in climatic change: a theory of the ice ages, *Meteorol. Monogr., 8*:37–62.
White, W. B., and T. P. Barnett, 1972: A servomechanism in the ocean/atmosphere system of the mid-latitude North Pacific, *J. Phys. Oceanog., 2*:372–381.
White, W. B., and A. E. Walker, 1973: Meridional atmospheric teleconnections over the North Pacific from 1950 to 1972, *Mon. Wea. Rev., 101*:817–822.
Wiin-Nielsen, A., 1972: Simulations of the annual variation of the zonally averaged state of the atmosphere, *Geofys. Publ., 28*:1–45.
Williams, G. P., and D. R. Davies, 1965: A mean motion model of the general circulation, *Q. J. R. Meteorol. Soc., 91*:471–489.
Williams, J., R. G. Barry, and W. M. Washington, 1973: Simulation of the climate at the last glacial maximum using the NCAR global circulation model, *J. Appl. Meteorol.* (to be published).
Willson, M. A. G., 1973: Statistical–dynamical modelling of the atmosphere, *Internal Scientific Report No. 17,* Commonwealth Meteorol. Res. Centre, Melbourne, 53 pp.

Wilson, A. T., 1964: Origin of ice ages: an ice shelf theory for Pleistocene glaciation, *Nature, 201*:147–149.

Wilson, C. L. (Chairman), 1970: Study of Critical Environmental Problems (SCEP) Report, *Man's Impact on the Global Environment*, W. H. Matthews, W. W. Kellogg, and G. D. Robinson, eds., MIT Press, Cambridge, Mass.

Wilson, C. L. (Chairman), 1971: Study of Man's Impact on Climate (SMIC) Report, *Inadvertent Climate Modification*, W. H. Matthews, W. W. Kellogg, and G. D. Robinson, eds., MIT Press, Cambridge, Mass., 308 pp.

Winstanley, D., 1973a: Recent rainfall trends in Africa, the Middle East and India, *Nature, 244*:464–465.

Winstanley, D., 1973b: Rainfall patterns and general atmospheric circulation, *Nature, 245*:190–194.

Wright, H. E., Jr., 1971: Lake Quaternary vegetational history of North America, in *The Late Cenozoic Glacial Ages*, K. Turekian, ed., Yale U. P., New Haven, Conn., pp. 425–464.

Wright, H. E., Jr., and D. G. Frey, eds., 1965: *The Quaternary of the United States*, Princeton U. P., Princeton, N.J., 922 pp.

Wyrtki, K., 1973: Teleconnections in the equatorial Pacific Ocean, *Science, 180*: 66–68.

Wyrtki, K., 1974: Sea level and the seasonal fluctuations of the equatorial currents in the western Pacific ocean, *J. Phys. Oceanog., 4*:91–103.

Yamamoto, G., and M. Tanaka, 1972: Increase of global albedo due to air pollution, *J. Atmos, Sci., 29*:1405–1412.

APPENDIX A
SURVEY OF PAST CLIMATES

INTRODUCTION

The earth's climates have always been changing, and the magnitude of these changes has varied from place to place and from time to time. In some places the yearly changes are so small as to be of minor interest, while in others the changes can be catastrophic, as when the monsoon fails or unseasonable rain delays the planting and harvesting of basic crops. On a longer time scale, certain decades have striking and anomalous characteristics, such as the severe droughts that affected the American Midwest during the 1870's, 1890's, and 1930's and the high temperatures recorded globally during the 1940's. And on still longer time scales, the climatic regimes that dominated certain centuries brought significant changes in the global patterns of temperature, rainfall, and snow accumulation. For example, northern hemisphere winter temperatures from the midfifteenth to the midnineteenth centuries were significantly lower than they are today. The late nineteenth century represented a period of transition between this cold interval—sometimes known as the Little Ice Age—and the thermal maximum of the 1940's. Some idea of the magnitude of the climatic changes that characterized the Little Ice Age can be gained from a study of proxy or natural records of climate, such as those of alpine glaciers. As shown in Figure A.1, as late as the midnineteenth century the termini of these glaciers were still advanced well beyond their present limits.

The practical as well as the purely scientific value of understanding

FIGURE A.1 The Argentière glacier in the French Alps. (a) An etching made about 1850, showing the extent of the glacier during the waning phase of the Little Ice Age. (b) Photograph of the same view taken in 1966. [From LeRoy Ladurie (1971).]

the processes that bring about climatic change is self-evident. Only by understanding the system can we hope to comprehend its past and to predict its future course. This objective can be achieved only by studying the workings of the global climate machine over a time span adequate to record a representative range of conditions in nature's own laboratory, and for this the record of past climates is indispensable.

From the evidence discussed below and summarized in Figure A.2 we conclude that a satisfactory perspective of the history of climate can be achieved only by the analysis of observations spanning the entire time range of climatic variation, say, from 10^{-1} to 10^9 years. Near the short end of this range there is a rich instrumental record to collate and analyze, although as discussed elsewhere in this report, awkward gaps exist in our knowledge of many parts of the air–sea–ice system during even the past hundred years. As the time scale of observations is lengthened to include earlier centuries, the direct instrumental record becomes less and less adequate. A continuous time series of observations as far back as the seventeenth century is available for only one area. For earlier times the instrumental record is blank, and indirect means must be found to reconstruct the history of climate.

The science of paleoclimatology is concerned with the earth's past climates, and that branch which seeks to map the reconstructed climates may be referred to as paleoclimatography. So defined, the science of paleoclimatology does far more than satisfy man's natural curiosity about the past; it provides the only source of direct evidence on processes that change global climate on time scales longer than a century. When calibrated and assembled into global arrays, these data will be essential in the reconstruction of paleoclimates with numerical models.

Nature of Paleoclimatic Evidence

The subject of ancient climates may conveniently be approached in terms of the nature of the climatic record, whether from human (historical) recordings or from proxy or natural climatic indicators. It is therefore convenient to identify historical climatic data and proxy climatic data as sources of paleoclimatic evidence.

Prior to the period of instrumental record, historical climatic data are found in books, manuscripts, logs, and other documentary sources and provide valuable (although fragmentary) climatic evidence before the advent of routine meteorological observations. Lamb (1969) has pioneered the collection of such data and has charted the main course of climate over Western Europe during the past 1000 years [Figure A.2(b)]. Where the historical or manuscript record overlaps the instru-

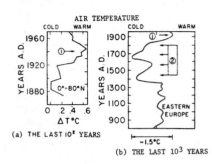
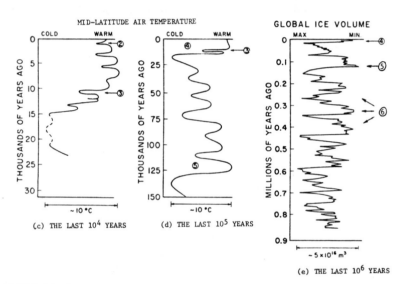

FIGURE A.2 Generalized trends in global climate: the past million years. (a) Changes in the five-year average surface temperatures over the region 0–80 °N during the last 100 years (Mitchell, 1963). (b) Winter severity index for eastern Europe during the last 1000 years (Lamb, 1969). (c) Generalized midlatitude northern hemisphere air-temperature trends during the last 15,000 years, based on changes in tree lines (LaMarche, 1974), marginal fluctuations in alpine and continental glaciers (Denton and Karlén, 1973), and shifts in vegetation patterns recorded in pollen spectra (van der Hammen et al., 1971). (d) Generalized northern hemisphere air-temperature trends during the last 100,000 years, based on midlatitude sea-surface temperature and pollen records and on worldwide sea-level records (see Figure A.13). (e) Fluctuations in global ice-volume during the last 1,000,000 years as recorded by changes in isotopic composition of fossil plankton in deep-sea core V28–238 (Shackleton and Opdyke, 1973). See legend for identification of symbols (1) through (6).

mental record, the climatic reconstructions may be confirmed and calibrated by the latter.

In contrast, the proxy record of climate makes use of various natural recording systems to carry the record of climate back into the past. Records from well-dated tree rings, annually layered (or varved) lake sediments, and ice cores resemble the historical data in that values can be associated with individual years and may be calibrated with modern data to extend the climatic record for many centuries, and in certain favored sites for as long as 8000 to 10,000 years. Other recording systems, such as the pollen concentration in lake sediments and fossil organisms and oxygen isotopes in ocean sediments, have less resolution but may provide continuous records extending over many tens of thousands of years. These and other characteristics of proxy climatic data sources are summarized in Table A.1.

In general, the older geological records provide only fragmentary and generally qualitative information but constitute our only records extending back many millions of years. For the past one million years, however, and especially for the past 100,000 years, the record is relatively continuous and can be made to yield quantitative estimates of the values of a number of significant climatic parameters. These include the total volume of glacial ice (and its inverse, the sea-level), the air temperature and precipitation over land, the sea-surface temperature and salinity for much of the world ocean, and the general trend of air temperature over the polar ice caps.

Like sensing systems made by man, each natural paleoclimatic indicator must be calibrated, and each has distinctive performance characteristics that must be understood if the data are to be interpreted correctly. In discussing these sources it is useful to distinguish between those paleoclimatic indicators that are more or less continuous recorders of climate, such as tree rings and varves, and those whose records are episodic, such as mountain glaciers. We should also consider the minimum attainable sampling interval that is characteristic of a particular paleoclimatic indicator (see Table A.1). Thus, tree rings, varves, and some ice cores can be sampled at intervals of one year, pollen or other sedimentary fossil samples only rarely represent less than about 100 years, and many geological series are sampled over intervals representing a thousand years or more. These figures reflect differences in the resolving power of each proxy indicator. Climate-induced changes in a plant community as reflected in pollen concentrations, for example, are relatively slow; the high-frequency information is lost, but low-frequency changes are preserved. In contrast, tree-ring

TABLE A.1 Characteristics of Paleoclimatic Data Sources

Proxy Data Source	Variable Measured	Continuity of Evidence	Potential Geographical Coverage	Period Open to Study (yr B.P.)	Minimum Sampling Interval (yr)	Usual Dating Accuracy (yr)	Climatic Inference
Layered ice cores	Oxygen isotope concentration, thickness (short cores)	Continuous	Antarctica, Greenland	10,000	1–10	±1–100	Temperature, accumulation
	Oxygen isotope concentration (long cores)	Continuous	Antarctica, Greenland	100,000+	Variable	Variable	Temperature
Tree rings	Ring-width anomaly, density, isotopic composition	Continuous	Midlatitude and high-latitude continents	1,000 (common) 8,000 (rare)	1	±1	Temperature, runoff, precipitation, soil moisture
Fossil pollen	Pollen-type concentration (varved core)	Continuous	Midlatitude continents	12,000	1–10	±10	Temperature, precipitation, soil moisture
	Pollen-type concentration (normal core)	Continuous	50 S to 70 N	12,000 (common) 200,000 (rare)	200	±5%	Temperature, precipitation, soil moisture
Mountain glaciers	Terminal positions	Episodic	45 S to 70 N	40,000	—	±5%	Extent of mountain glaciers

Source	Feature	Continuity	Location			Variable	Area of ice sheets
Ice sheets	Terminal positions	Episodic	Midlatitude to high latitudes	25,000 (common) 1,000,000 (rare)	—		Temperature, precipitation, drainage
Ancient soils	Soil type	Episodic	Lower and mid-latitudes	1,000,000	200	±5%	Evaporation, runoff, precipitation, temperature
Closed-basin lakes	Lake level	Episodic	Midlatitudes	50,000	1–100 (variable)	±5%	Temperature, precipitation
Lake sediments	Varve thickness	Continuous	Midlatitudes	5,000	1	±1	Wind direction
Ocean sediments (common deep-sea cores, 2–5 cm/1000 yr)	Ash and sand accumulation rates	Continuous	Global ocean (outside red clay areas)	200,000	500+	±5%	
	Fossil plankton composition	Continuous	Global ocean (outside red clay areas)	200,000	500+	±5%	Sea-surface temperature, surface salinity, sea-ice extent
	Isotopic composition of planktonic fossils; benthic fossils; mineralogic composition	Continuous	Global ocean (above $CaCO_3$ compensation level)	200,000	500+	±5%	Surface temperature, global ice volume; bottom temperature and bottom water flux; bottom water chemistry
(rare cores, >10 cm/1000 yr)	As above	Continuous	Along continental margins	10,000+	20	±5%	As above
(cores, <2 cm/1000 yr)	As above	Continuous	Global ocean	1,000,000+	1000+	±5%	As above
Marine shorelines	Coastal features, reef growth	Episodic	Stable coasts, oceanic islands	400,000	—	±5%	Sea level, ice volume

records and isotopic records in ice cores respond yearly and even seasonally in favored sites.

Each proxy record also has a characteristic chronologic and geographic range over which it can be used effectively. Tree-ring records go back several thousand years at a number of widely distributed continental sites, pollen records have the potential of providing synoptic coverage over the continents for the past 12,000 years or so, and a nearly complete record of the fluctuating margins of the continental ice sheets is available for about the last 40,000 years. Planktonic and benthic fossils from deep-sea cores can in principle provide nearly global coverage of the ocean going back tens of millions of years, although sampling difficulties have thus far limited our access to sediments deposited during the last several hundred thousand years.

Although instrumental records best provide the framework necessary for the quantitative understanding of the physical mechanisms of climate and climatic variation with the aid of dynamical models, the increasingly quantitative and synoptic nature of paleoclimatic data will add a much-needed perspective. As discussed elsewhere in this report, it is therefore important that the historical and proxy records of past climate be systematically assembled and analyzed, in order to provide the data necessary for a satisfactory description of the earth's climates.

Instrumental and Historical Methods of Climate Reconstruction

Over the past three centuries, the development of meteorological instruments and appropriate "platforms" for sensing the state of the atmosphere–hydrosphere–cryosphere system has produced an important storehouse of quantitative information pertaining to the earth's climates. Time series of these records show that climate undergoes considerable variation from year to year, decade to decade, and century to century. From a practical viewpoint, much of this information mirrors the economically more important climatic variations, as found, for example, in the changes of crop and animal production, the patterns of natural flora and fauna, and the variations of the levels of lakes and streams and the extent of ice. Generally the term "climate" is understood to describe in some fashion the "average" of such variations. As discussed in Chapter 3, a complete description of a climatic state would also include the variance and extremes of atmospheric behavior, as well as the values of all parameters and boundary conditions regarded as external to the climatic system.

In discussing the reconstruction of climates from instrumental data, several characteristics of past and present observational systems should

APPENDIX A 135

be considered. First, instrumental observations have been obtained for the most part for the purpose of describing and forecasting the weather. Hence, although extensive records of such weather elements as temperature, precipitation, cloudiness, wind, and observations are available, they are inadequate for many climatic purposes. There exist few direct measurements related to the thermal forcing functions of the atmosphere–hydrosphere–cryosphere system, such as the solar constant; the radiation, heat, and moisture budgets over land and ocean surfaces; the vegetative cover; the distribution of snow and ice; the thermal structure of the oceanic surface layer; and atmospheric composition and turbidity.

Second, observational records may be expected to contain errors due to changes in instrument design and calibration and to changes in instrument exposure and location. There is therefore a need to establish and maintain conventional observations at reference climatological stations and a need to identify, insofar as possible, "benchmark" records of past climate. Such observations are needed to supplement the climatic monitoring program described elsewhere (see Chapter 6).

Third, the time interval over which portions of the climatic system need to be described are very different. If, for example, the fluctuations in the volume and extent of the polar ice caps are to be studied, a time interval of order 100,000 years (the maximum residence time of water in the ice caps) is required. Or if atmospheric interaction with the deep oceans is to be considered, then a time interval of the order 500 years (the residence time of bottom water) is required. It is therefore apparent that the period of instrumental records covering the past century or two is long enough only to have sampled a portion of such climatic responses, and that our information on older climates must come from historical sources and from the various natural (proxy) indicators of climate described earlier. Although such records will always be fragmentary, we should recognize their unique value in describing the past behavior of the earth's climatic system.

For practical reasons, it has been convenient to compute climatic statistics over relatively short intervals of time, such as 10, 20, or 30 years, and to designate the 30-year statistics as climatic "normals." It is important to note, however, that the most widely accepted climatic "normals" (for the period 1931–1960) represent one of the most *abnormal* 30-year periods in the last thousand years (Bryson and Hare, 1973). As noted elsewhere in this report, the entire last 10,000 years are themselves also abnormal in the sense that such (interglacial) climates are typical of only about one tenth of the climatic record of the last million years.

While continuous observations of atmospheric pressure, temperature, and precipitation are available at a few locations from the late seventeenth century, such as the record of temperature in Central England assembled by Manley (1959), it is only since the early part of the eighteenth century that the spatial coverage of observing stations has permitted the mapping of climatic variables on even a limited regional scale. These and other scattered early observations of rainfall, wind direction, and sea-surface temperature have been summarized by Lamb (1969). Only since about 1850 are reliable decadal averages of surface pressure available for most of Europe, and only since about 1900 are there reliable analyses for the midlatitudes of the northern hemisphere, as shown in Figure A.3. And only since about 1950 does the surface observational network begin to approach adequate coverage over the continents; large portions of the oceans, particularly in the southern hemisphere, remain inadequately observed.

For the climate of the free atmosphere, the international radiosonde network permits reliable analyses for the midlatitudes of the northern hemisphere only since the 1950's, and less than adequate coverage exists over the rest of the globe. Beginning in the 1960's, routine observations from satellite platforms have begun to make possible global observations of a number of climatic variables, such as cloudiness, the planetary albedo, and the planetary heat budget. Yet many important quantities, such as the heat and moisture budgets at the earth's surface and the thermal structure and motions of the oceanic surface layer, remain largely unobserved on even a local scale.

Biological and Geological Methods of Climate Reconstruction

During the first three decades of the nineteenth century, Venetz in Switzerland and Esmark in Norway inferred the existence of a prehistoric ice age from the study of vegetation-covered moraines and other glacial features in the lower reaches of mountain valleys. After a century of effort, the literature of paleoclimatology has become so diverse, and so burdened by stratigraphic terminology, that it is useful to provide a summary of paleoclimatic techniques.

The quantitative description of past climates as determined by biological and geological records requires the development of paleoclimatic monitoring techniques and the construction of time scales by suitable chronometric or dating methods. In general, the second of these problems is the more difficult.

Beyond the range of ^{14}C dating (the past 40,000 years), it is only since about 1970 that the main chronology of the climate of the past 100,000 years has become clear; and only since 1973 that the main features of the chronology of the past million years have been estab-

APPENDIX A

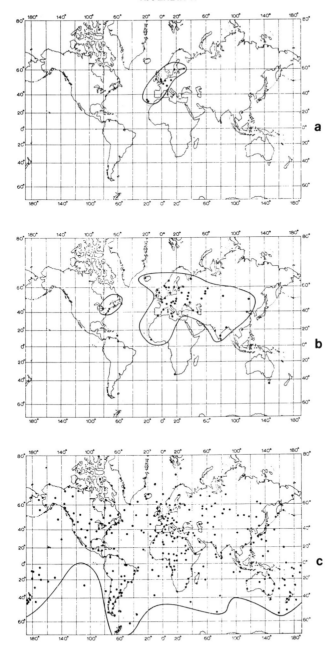

FIGURE A.3 Growth of the network of surface pressure observations and of the area that can be covered by reliable 10-year average isobars (Lamb, 1969). (a) 1750–1759, (b) 1850–1859, (c) 1950–1959.

lished. Key discoveries in these time ranges have been in the sea-level records of oceanic islands and in the sedimentary records of deep-sea cores. In preparing this survey, the chronology of these records has been used as a framework into which the data from more fragmentary or poorly dated records have been fitted.

Monitoring Techniques

The problem of developing a paleoclimatic monitoring technique—or finding something meaningful to plot—may be broken down into three subproblems. A natural climatic record must be (a) identified, (b) calibrated, and (c) obtained from a stable recording medium.

Identification of Natural Climatic Records A number of different monitoring techniques that can provide data for paleoclimatic inference are summarized in Table A.1 and are based on observations of fossil pollen, ancient soil types, lake deposits, marine shore lines, deep-sea sediments, tree rings, and ice sheets and mountain glaciers. The techniques that are emphasized here are those that in general yield more or less continuous time series. Other types of proxy data are also useful in the reconstruction of climatic history (see, for example, Flint, 1971, or Washburn, 1973).

Calibration of Paleoclimatic Records Many proxy records must be calibrated to provide an estimate of the climatic parameter of interest. The elevation of an ancient coral reef, for example, is a record of a previous sea level; but before it can be used for paleoclimatic purposes the effect of local crustal uplift or subsidence must be removed (Bloom, 1971; Matthews, 1973; Walcott, 1972).

Another example may be cited from paleontology, where the taxonomic composition of fossil assemblages and the width of tree rings are known to reflect the joint influence of several ecological and environmental factors of climatic interest. Here appropriate statistical techniques are used to define indices that give estimates of the individual paleoclimatic parameters, such as air temperature, rainfall, or sea-surface temperature and salinity. In the case of tree rings, although each tree responds only to the local temperature, moisture, and sunlight, for example, by averaging over many sites, the trees' response may be related to the large-scale distribution of rainfall and surface temperature. In this way a statistical relationship may be established with a variety of parameters, even though they may not be direct causes of tree growth. When such tree-ring data are carefully dated they can thus provide estimates of the past regional variations of climatic elements such as precipitation, temperature, pressure, drought, and stream flow (Fritts et al., 1971). These methods yield what are called transfer functions, which serve to transform one set of time-varying signals to another set that represents the desired paleoclimatic estimates. In addition to their application to tree-ring data, multivariate statistical-analysis techniques have been successfully applied to marine fossil data (Imbrie and Kipp, 1971; Imbrie, 1972; Imbrie et al., 1973) and to fossil pollen data (Bryson et al., 1970; Webb and Bryson, 1972). Typical results indicate, for example, that average winter sea-surface temperatures 18,000 years ago in the Caribbean were about 3°C lower than today, while those in midlatitudes of the North Atlantic were about 10°C below present levels.

The oxygen isotope ratio $^{18}O/^{16}O$ as it is preserved in different materials is used in three separate paleoclimatic monitoring techniques. Although the results are interpreted differently, in each technique the ratio is measured as the departure $\delta^{18}O$ from a standard, with positive values indicating an excess of the heavy isotope. One technique examines the ratio in polar ice caps, where the values of $\delta^{18}O$ are generally on the order of 30 parts per thousand lower than in the oceanic reservoir, because of the precipitation and isotopic enrichment that accompanies the transport of water vapor into high latitudes. As shown by Dansgaard (1954) and by Dansgaard et al. (1971) the value of $\delta^{18}O$ in each accumulating layer of ice is closely related to the temperature at which precipitation occurs over the ice. Although complicating effects make it impossible to convert the $\delta^{18}O$ curve into an absolute measure of air temperature, the isotopic time series are extraordinarily detailed.

Another isotopic technique records $\delta^{18}O$ in the carbonate skeletons of planktonic marine fossils (Emiliani, 1955, 1968). Here the ratio is determined by the isotopic ratio and temperature in the near-surface water in which the organisms live. Work by Shackleton and Opdyke (1973) demonstrates that the observed ratio is predominantly influenced by the isotopic ratio in the seawater. Hence the isotopic curve reflects primarily the changing volume of polar ice, which, upon melting, releases isotopically light water into the ocean.

A third technique measures the isotopic ratio in benthic fossils whose skeletons reflect conditions prevailing in bottom waters. By making the assumption that the temperature of bottom water underwent little change over the past million years, the difference between the isotopic ratio observed in benthic and planktonic fossils can be used to estimate changes in surface-water temperatures. Initial application of this technique (Shackleton and Opdyke, 1973) provides an independent confirmation of the previously cited estimate of glacial-age Caribbean temperatures obtained by paleontological techniques. Over time spans on the order of tens of millions of years, measurements of $\delta^{18}O$ in benthic fossils offer a means of tracing changes in bottom water in which the effects of changing polar temperatures and ice volumes are combined (Douglas and Savin, 1973).

Evaluation of the Recording Medium All paleoclimatic techniques require that ambient values of a climatic parameter be preserved within individual layers of a slowly accumulating natural deposit. Such deposits include sediments left by melting glaciers on land; sediments accumulating in peat bogs, lakes, and on the ocean bottom; soil layers; layers accumulating in polar ice caps; and the annual layers of wood formed in growing trees. Ideally, a recording site selected for paleoclimatic work should yield long, continuous, and evenly spaced time series. The degree to which these qualities are realized varies from site to site, so that distortions and nonuniformities in each record must be identified and removed. The stratigraphic techniques by which this screening is accomplished will not be discussed here, although the reader should be aware that (with the exception of tree rings) some degree of chronological distortion will occur in all paleoclimatic curves where chronometric control is lacking.

To enable the reader to form his own judgments as to the chronology of past climatic changes, most of the paleoclimatic curves given in this report show explicitly the time control points between which the data are spaced in proportion to their relative position in the original sedimentary record. This procedure assumes that accumulation was constant between the time controls, which is a reasonable assumption in favorable environments. In other cases this assumption introduces a distortion in the signal and a consequent uncertainty in the timing of the inferred climatic variations.

Each of the recording media used in paleoclimatography has characteristic limitations and advantages. As summarized in Table A.1, the reconstruction of past climates requires evidence from a variety of techniques, each yielding time series of different lengths and sampling intervals and reflecting variations in different regions. The tree-ring record, for example, provides evenly spaced and continuous annual records, but only for the past few thousand years. The ice-margin record of both valley and continental glaciers is discontinuous, especially prior to about 20,000 years ago, because each major glacial advance tends to obliterate (or at least to conceal) the earlier evidence. Records of lake levels and sea levels are also discontinuous. The former rarely extend back more than 50,000 years, although the latter extend back several hundred thousand years. Soil sequences display great variability in sedimentation rate but provide continuous climatic information for sites on the continents where other records are not available (or are discontinuous); in favored sites, the soil record extends back about a million years. Pollen records are usually continuous but are rarely longer than 12,000 years. Deep-sea cores provide material for the study of fossils, oxygen isotopes, and sedimentary chemistry. These records are relatively continuous over the past several hundred thousand years and are distributed over large parts of the world ocean. Their relatively uniform but low deposition rates, however, generally limit the chronological detail obtainable. Cores taken in the continental ice sheets provide a detailed and generally continuous record for many thousands of years, although their interpretation is handicapped by the lack of fully adequate models of the ice flow with its characteristic velocity–temperature feedback.

Chronometric Techniques

The problem of constructing a paleoclimatic chronology has been approached by four direct methods and one indirect method.

Dendrochronology The most accurate direct dating is achieved in tree-ring analysis, in which many records with overlapping sets of rings are matched. With sufficient samples, virtual certainty in the dates of each annual layer may be obtained, and a year-by-year chronology can be established for periods covered by the growth records of both living and fossil trees. Such records are especially valuable for studying variations of climate during the last few hundred years and can be extended to many of the land regions of the world.

Analysis of Annually Layered Sediments In favored locations, lakes with annually layered bottom sediments provide nearly the same time control as do tree rings. Some ice cores and certain marine sediment cores from regions of high deposition rates also contain distinct annual layers. These data, along with tree rings and historical records, are the only source of information on the high-frequency portion of the spectrum of climatic variation.

Radiocarbon Dating The advent of the ^{14}C method in the early 1950's was a major breakthrough in paleoclimatography, for it made possible the development of a reasonably accurate absolute chronology of the past 40,000 years in widely distributed regions. Prior knowledge was essentially limited to dated tree-ring sequences (for the past several thousand years) and to varve-counted sequences in Scandinavia (extending back to about 12,000 years). The ^{14}C method has an accuracy of about ±5 percent of the age being determined; that is, material 10,000 years old could be dated within the range 9500–10,500 years. The calibration of ^{14}C ages against those determined from dendrochronology gives insight into the variations of atmospheric ^{14}C production rates over the past 7000 years (Suess, 1970).

Decay of Long-Lived Radioactivities   These methods employ daughter products of uranium decay or the production of ^{40}Ar through potassium decay. Used under favorable circumstances, one of the uranium methods (the decay of ^{230}Th) can provide approximate average sedimentation rates in deep-sea cores. The other method (the growth of ^{230}Th) can be used successfully on fossil corals to provide discrete dates for shoreline features recording ancient sea levels. Together, these techniques have provided a reasonably satisfactory chronology of the past 200,000 years with a dating accuracy of about ±10 percent. Our chronology for older climatic records is based on the well-known K/Ar technique, applied to terrestrial lava flows and ash beds. This technique has provided, for example, the important dates for paleomagnetic reversal boundaries.

Stratigraphic Correlation with Dated Sequences   Much of the absolute chronology of climatic sequences is supplied by an indirect method, namely, the stratigraphic correlation of specific levels in an undated sequence with dated sequences from another location. For example, a particular glacial moraine that lacks material for ^{14}C dating may be identified with another formed at the same time that has datable material. Such correlation by direct physical means is limited to relatively small regions, however, and stratigraphic correlation techniques must be used. Three such methods form the backbone of the chronology of paleoclimate: biostratigraphy, isotope stratigraphy, and paleomagnetic stratigraphy.

The techniques of biostratigraphy use the levels of extinction or origin of selected species as the basis for correlation. This method has enabled Berggren (1972), for example, to devise a time scale of the past 65,000,000 years that is widely used as a basis for historical interpretation. Isotope stratigraphy, applicable only to the marine realm, makes use of the fact that the record of oxygen isotope variations—

which reflects chiefly the global ice volume—has distinctive characteristics that permit the correlation of previously undated sequences. The application of paleomagnetic correlation techniques has revolutionized our approach to the climatic history of the past several million years. Their importance stems from the fact that the principal magnetic reversal boundaries, which have occurred irregularly about every 400,000 years, are recorded in both marine and continental sedimentary sequences.

Regularities in Climatic Series

On the assumption that climatic changes are more than just random fluctuations, paleoclimatologists have long sought evidence of regularities in proxy records of the earth's climatic history. Many have found what they believe to be firm evidence of order and refer to the chronological patterns as "cycles." Although the number of records is limited, and hard statistical evidence is sometimes lacking, it is nevertheless convenient to describe some of the larger climatic changes in terms of quasi-periodic fluctuations or cycles with specified mean wavelengths or periods, in the sense that they describe the apparent repetitive tendency of certain sequences of climatic events. For example, many aspects of the global ice fluctuations during the last 700,000 years may be summarized in terms of a 100,000-year cycle [see Figure A.2(e)]. Each such period is marked by a gradual transition from a relatively ice-free climate (or interglacial) to a short, intense glacial maxima and followed by an abrupt return to ice-free climate. No two such cycles are the same in detail, however, and should not be construed as indicating strict periodicities in climate.

Some paleoclimatic cycles may be periodic, or at least quasi-periodic, and rest on evidence that is exclusively or mainly chronological. The best example is the approximate 100,000-year cycle found from the spectral analysis of time series, such as that shown in Figure A.4. For the 100,000-year cycle, as well as some of the higher-frequency fluctuations that modify it, there is circumstantial evidence to suggest that these have in some way been induced by secular variations of the earth's orbital parameters, which are known to alter the latitudinal pattern of the seasonal and annual solar radiation received at the top of the atmosphere. For the 2500-year (and shorter) fluctuations suggested by some proxy data series, the causal mechanism is unknown.

With the possible exception of the approximately 100,000-year quasi-periodic fluctuation referred to above, the quasi-biennial oscillation (of

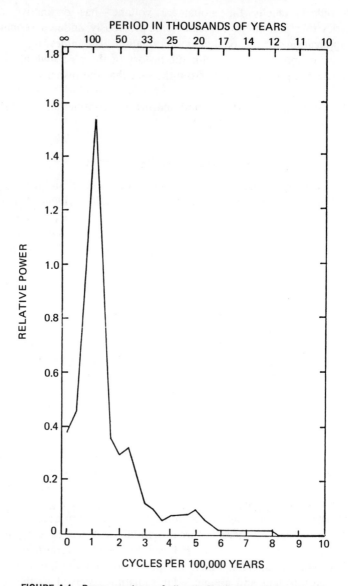

FIGURE A.4 Power spectrum of climatic fluctuations during the last 600,000 years according to Imbrie and Shackleton (1974). The data analyzed are time-series observations of $\delta^{18}O$ in fossil plankton in the upper portion of a deep-sea core in the equatorial Pacific, interpolated at intervals of 2500 years (Shackleton and Opdyke, 1973). This ratio reflects fluctuations in global ice-volume.

2–3 year period) is the only quasi-periodic oscillation whose statistical significance has been clearly demonstrated. This is not to say that other such fluctuations in climate are absent but rather that much further analysis of proxy records is required. A question of equal importance is the shape of the continuum variance spectrum of climatic fluctuations. A uniform distribution of variance as a function of frequency (or "white noise") would imply a lack of predictability in the statistical sense or a lack of "memory" of prior climatic states. A "red-noise" spectrum, on the other hand, in which the variance decreases with increasing frequency, implies some predictability in the sense that successive climatic states are correlated. The existence of nonzero autocorrelations in such a spectrum implies that some portion of the climatic system retains a "memory" of prior states. In view of the relatively short memory of the atmosphere, it seems likely that this is provided by the oceans on time scales of years to centuries and by the world's major ice sheets on longer times scales.

An initial estimate of the variance spectrum of temperature has been made from the fluctuations on time scales from 1 to 10,000 years by Kutzbach and Bryson (1974) and is shown in Figure A.5. This spectrum has been constructed from a combination of calibrated botanical, chemical, and historical records, along with instrumental records in the North Atlantic sector. As may be seen in Figure A.5(a), the variance spectral density increases with decreasing frequency (increasing period) over the entire frequency domain but is most pronounced for periods longer than about 30 years. In Figure A.5(b), the spectrum of the same time series is shown with frequency on a logarithmic scale and the ordinate as spectral density (V) times frequency (f), so that equal areas represent equal variance. Again, for periods longer than about 30–50 years, the observed temperature spectrum is seen to depart significantly from the white-noise continuum associated with the high-frequency portion of the spectrum. The determination of the character of the variance spectrum of the various climatic elements remains largely a task for the future.

We will use the term "cycle" in the following paragraphs to designate such quasi-periodic sequences of climatic events, since there appears to be no other word or phrase that conveys the concept of a series of generally similar events spaced at reasonably regular intervals in time. Although our knowledge of the record of past climates has improved greatly during the last decade, a much broader paleoclimatic data base is clearly required. Only then can adequate spectral analyses be performed and the spatial and temporal structure of paleoclimatic variations firmly established.

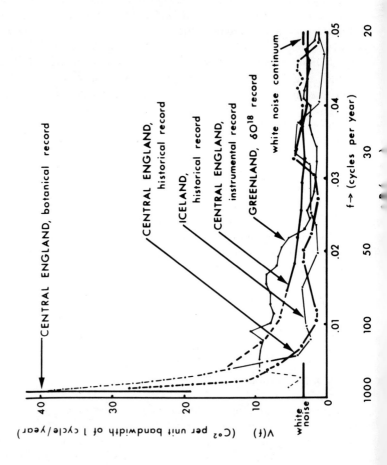

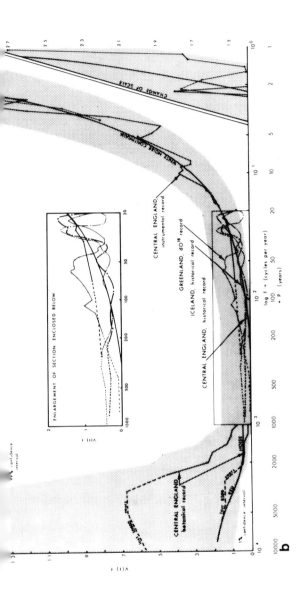

FIGURE A.5 (a) Composite variance spectrum of temperature on time scales of 10 to 10^5 years derived from instrumental, historical, botanical, and oxygen isotope ($\delta^{18}O$) records. Ordinate is variance $V(f)$ in °C^2 per unit frequency bandwidth of 1 cycle per year, and abscissa is a linear frequency scale. The sources of data are Central England botanical and historical records (Lamb et al., 1966), Central England instrumental records (Manley, 1959), Iceland historical records (Bergthorsson, 1962, as reproduced in Bryson and Hare, 1973), and Greenland $\delta^{18}O$ records (Dansgaard et al., 1971). From Kutzbach and Bryson (1974). (b) Composite variance spectrum at temperature on time scales of 1 to 10^4 years derived from instrumental, historical, botanical, and oxygen isotope ($\delta^{18}O$) records. The ordinate is $V(f)$ times f in °C^2, and the abscissa is a logarithmic frequency scale. The four lowest frequency spectral estimates of each individual spectrum are connected by dashed lines to indicate that they are statistically unreliable. Shading indicates the generalized 1 percent and 99 percent confidence limits. The insert is an enlarged version of the intermediate-frequency range (from Kutzbach and Bryson, 1974).

CHRONOLOGY OF GLOBAL CLIMATE

Period of Instrumental Observations

A variety of meteorological indices have been used to characterize the climate and its temporal variations during the past century or more of extensive observations. Global- or hemisphere-averaged indices such as the surface temperature index shown in Figure A.6 are often used for this purpose. This index clearly suggests a worldwide warming beginning in the 1880's, followed by a cooling since the 1940's. The warming may be recognized as the last part of a complex but recognizable trend that has persisted since the end of the seventeenth century [see Figure A.2(b)].

The geographic patterns of temperature change during these overall warming and cooling epochs show considerable variability, with the largest changes concentrated in the polar regions of the northern hemisphere. Mitchell (1963) has shown that the pattern of temperature change during recent decades is consistent with concomitant changes in the large-scale atmospheric circulation as reflected in sea-level pressure. Less attention has been given to the more complex relationships between circulation variation and changing precipitation patterns, although Kraus (1955a, 1955b) and Lamb (1969) have considered this aspect of the problem.

Lamb and Johnson (1959, 1961, 1966) and Lamb (1969) have made an extensive analysis of certain features of atmospheric circulation

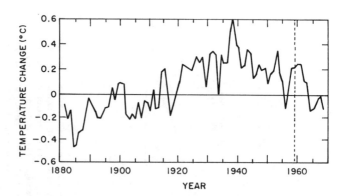

FIGURE A.6 Recorded changes of annual mean temperature of the northern hemisphere as given by Budyko (1969) and as updated after 1959 by H. Asakura of the Japan Meteorological Agency (unpublished results).

based on the observed and historically reconstructed surface pressure maps for individual months since about 1750. They have extracted such indices as the strength of the zonal and meridional flow, the position and wavelength of trough–ridge patterns, and the position and strength of subtropical pressure systems. The year-to-year and decade-to-decade changes in these indices reflect changing large-scale circulation patterns, which in turn are associated with changing patterns of temperature and precipitation. From the instrumental era for the North Atlantic sector, the typical variability over 20- to 30-year intervals of the low-level westerlies is $\pm 1-2$ m sec^{-1}, and that for the planetary-scale circulation features (such as large-scale troughs and ridges) is $\pm 1-2$ deg latitude and $\pm 10-20$ deg longitude.

Although changes in the position, pattern, and intensity of the general circulation are interrelated, such empirical studies suggest that longitudinal shifts have the most significant effects on the climatological temperature and precipitation patterns, at least for middle and higher latitudes. Examples of such shifts are shown in Figure A.7. In tropical and subtropical latitudes, on the other hand, latitudinal shifts appear to be more closely related to regional climatic variations, as indicated by the data of Figure A.8.

The Last 1000 Years

To obtain an indication of the climate in the northern hemisphere for the last 1000 years, Lamb (1969) has compiled manuscript references on the character of European weather and has developed an index of winter severity, as shown in Figure A.2(b) and A.9. Although different longitudes show somewhat different results, the trends shown by this index (the excess number of unusually mild or unusually cold winter months over months of opposite character) for the period since about 1700 have been validated by comparison with thermometer records. Other portions of the record have been cross-checked with data on glacial fluctuations, oxygen isotope variations, and tree growth, so that the main characteristics of European climate during this period are reasonably well known. LaMarche (1974) has constructed temperature and moisture records from the ring-width variations in trees at high-altitude arid sites in California [see Figure A.9(c)]. Comparisons of his data with those from Europe shown in Figure A.9(d) indicate a degree of synchrony in the major fluctuations of temperature between the west coast of North America and Western Europe during the last 1000 years.

The early part of the last millenium (about A.D. 1100 to 1400) is sometimes called the Middle Ages warm epoch but was evidently not as

150 UNDERSTANDING CLIMATIC CHANGE

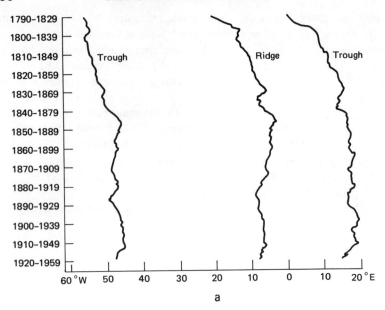

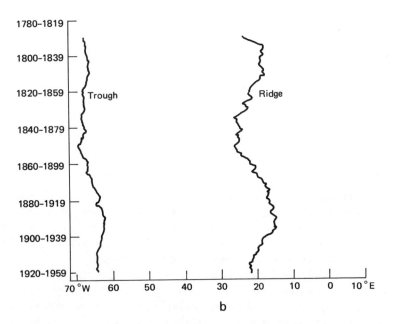

FIGURE A.7 Forty-year running means of the longitudes of the semipermanent surface pressure troughs and ridges in the North Atlantic (Lamb, 1969). (a) at 45 °N in January; (b) at 55 °N in July.

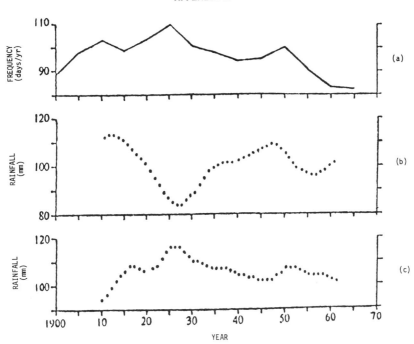

FIGURE A.8 Twenty-year running means of selected climatic indices (Winstanley, 1973). (a) Frequency (days per year) of westerly weather type over the British Isles (from Lamb, 1969); (b) winter-spring rainfall (mm) at 14 stations in North Africa and the Middle East; (c) summer monsoon rainfall (mm) at eight stations in the Sahel of North Africa and northwest India.

warm as the first half of the twentieth century. The period from about 1430 to 1850 is commonly known as the Little Ice Age, and some records indicate that this period had cold maxima in the fifteenth and seventeenth centuries. From such evidence we infer that the atmospheric circulation may have been more meridional than at present and characterized in western Europe and western North America by short, wet summers and long, severe winters.

During the Little Ice Age many glaciers in Alaska, Scandinavia, and the Alps advanced close to their maximum positions since the last major ice age thousands of years ago. A visual impression of these events in the French Alps was shown in Figure A.1. The expansion of the Arctic pack ice into North Atlantic waters caused the Norse colony in southwest Greenland to become isolated and perish; and in Iceland, grain that had grown for centuries could no longer survive.

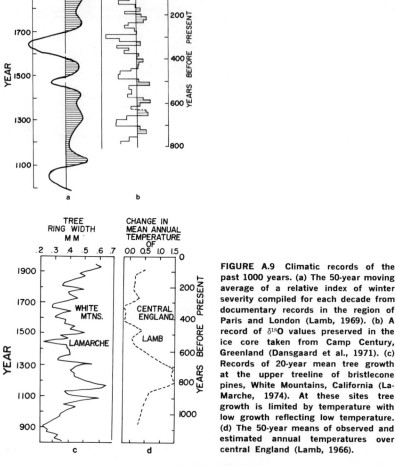

FIGURE A.9 Climatic records of the past 1000 years. (a) The 50-year moving average of a relative index of winter severity compiled for each decade from documentary records in the region of Paris and London (Lamb, 1969). (b) A record of $\delta^{18}O$ values preserved in the ice core taken from Camp Century, Greenland (Dansgaard et al., 1971). (c) Records of 20-year mean tree growth at the upper treeline of bristlecone pines, White Mountains, California (LaMarche, 1974). At these sites tree growth is limited by temperature with low growth reflecting low temperature. (d) The 50-year means of observed and estimated annual temperatures over central England (Lamb, 1966).

The Last 5000 Years

As indicated in Figure A.2(c), the period from 7000 to 5000 years ago was marked by temperatures warmer than those that prevail today [and is thus sometimes known as the hypsithermal interval (Flint, 1971)]. The last 5000 years is characterized by generally declining temperatures and a trend toward more extensive mountain glaciation (but not ice sheets) in all parts of the world (Porter and Denton, 1967). Close

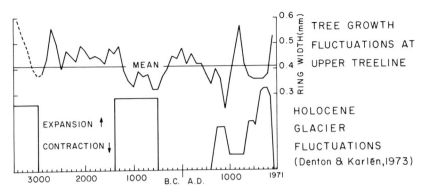

FIGURE A.10 Climatic records of the past 5000 years. (a) Average (100-year mean) ring widths of bristlecone pine at the upper treeline in the White Mountains of California (LaMarche, 1974). Positive growth departures indicate warm-season (April–October) temperatures above the long-term mean, with a total temperature range of about 4°F. (b) Records of the advance and retreat of Holocene Alaskan glaciers (Denton and Karlén, 1973).

examination of the records of mountain glaciers, treelines, and tree rings suggests that this general cooling trend was itself punctuated in many parts of the world by cold intervals centered at about 5300, 2800, and 350 years ago, as shown in Figure A.10. Much further analysis of proxy climatic records during this period is needed, including the evidence available from historical sources.

The Last 25,000 Years

The climatic record of the last 25,000 years is largely concerned with the present interglacial interval (or Holocene) and the terminal phases of the last major glaciation [see Figure A.2(d)]. Although the maximum ice extent occurred between about 22,000 and 14,000 years ago (see Figure A.11) the curves of ice accumulation and decline are not identical for the various ice sheets. The Laurentide ice sheet (which covered parts of eastern North America) and the Scandinavian ice sheet (which covered parts of northern Europe) reached their maximum extent between 22,000 and 18,000 years ago, while the Cordilleran ice sheet achieved its maximum only 14,000 years ago. The maximum areas of the northern hemisphere ice sheets during the past 25,000 years were about 90 percent of the maxima during the last million years of the Pleistocene (see Table A.2).

Widespread deglaciation began rather abruptly about 14,000 years

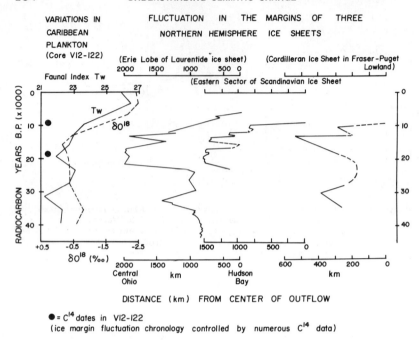

FIGURE A.11 Climatic records of the past 40,000 years. (a) Fluctuations in Caribbean plankton (core V12–122) interpreted as a record of sea-surface temperature in °C. (b) Fluctuations in the isotopic composition of Caribbean plankton interpreted as a record of changing global ice-volume. Both records are from Imbrie et al. (1973). Curves (c), (d), and (e) are time-distance plots of changes in the margins of three northern hemisphere ice sheets. Curve (c) is from Dreimanis and Karrow (1972), curves (d) and (e) are due to G. H. Denton, University of Maine (unpublished). The chronology of curves (a) and (b) is controlled by ^{14}C dates shown by solid circles; the ice-margin curves are controlled by numerous ^{14}C dates.

ago, and the waning phases of the continental ice sheets were characterized by substantial marginal fluctuations (Dreimanis and Karrow, 1972), as shown in Figure A.11. The Cordilleran ice sheet, which had just attained its maximum extent, melted rapidly and was gone by 10,000 years ago. The Scandinavian ice sheet lasted only slightly longer and retreated at the rate of about 1 km per year between about 10,000 and 9000 years ago. The climatic instability suggested by these fluctuations in the margins of the northern hemisphere's major ice sheets is corroborated by the records from fossil pollen, deep-sea cores, ice cores, and sea-level variations, as shown in Figure A.12, and by lacustrine records in western North America and Africa. By 8500 years ago the ice conditions in Europe had reached essentially their present

TABLE A.2 Characteristics of Existing Ice Sheets and of the Maximum Quaternary Ice Cover [a]

	Area (10^{12} m^2)	
Existing Glaciers		
Greenland	1.80	
Spitsbergen+Iceland	0.07	
Canadian Archipelago	0.15	
North America	0.08	
Europe+Asia	0.17	
South America	0.03	
Antarctica	12.59	
TOTAL AREA	14.99	(3% of earth's surface)
TOTAL ICE VOLUME [b]	2.5×10^7 km^3	
EQUIVALENT SEA-LEVEL CHANGE	70 m	
Maximum Quaternary Glaciation		
Greenland	2.30	
Spitsbergen+Iceland	0.44	
Alaska	1.03	
Cordillera	1.58	
Laurentide	13.39	
Scandinavia	6.67	
Europe	0.09	
Asia	3.95	
South America	0.87	
Antarctica	13.81	
Other	0.04	
TOTAL AREA	44.17	(9% of earth's surface)
TOTAL ICE VOLUME [b]	7.5×10^7 km^3	
EQUIVALENT SEA-LEVEL CHANGE	210 m	

[a] From Flint (1971).
[b] Based on the present ice thickness of 1700 m in Greenland and Antarctica.

state, and in North America the ice sheets had shrunk to about their present extent by about 7000 years ago.

How widespread and synchronous these fluctuations were is not yet known, but evidence is growing that there were several periods of widespread cooling and glacial expansion in the regions bordering the Atlantic Ocean [see Figure A.2(c)], spaced about 2500 years apart. One of these glacial advances (the Younger Dryas event, about 10,800 to 10,100 years ago) was a climatic event of unparalleled abruptness in Europe, establishing itself within a century or less and lasting for some 700 years. Northern forests that had advanced during the pre-

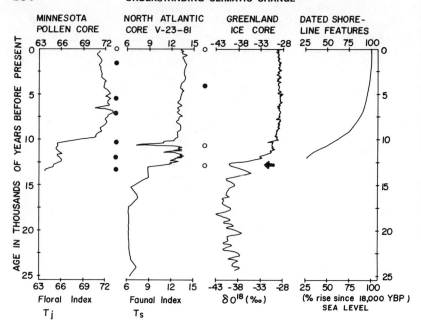

FIGURE A.12 Climatic records of the last 25,000 years. (a) A floral index reflecting changes in vegetation in Minnesota, as documented by pollen counts in a bog core (Webb and Bryson, 1972). The index is an estimate of July air temperature in °F. (b) A faunal index reflecting changes in foraminiferal plankton in a core west of Ireland, from C. Sancetta, Brown University (unpublished). The index is an estimate of August sea-surface temperature in °C. (c) Values of $\delta^{18}O$ in the ice-core of Camp Century, Greenland (Dansgaard et al., 1971). The isotope ratio is judged to reflect air-temperature variations over the ice cap, with the more negative values associated with colder temperatures. (d) Generalized curve of numerous sea-level records (Bloom, 1971). Chronology of curves (a) and (b) is established by ^{14}C dates (solid circles) and stratigraphic correlation with ^{14}C dates (open circles). Chronology for curve (c) above arrow (12,700 years ago) taken from Dansgaard et al., (1971); below arrow, the chronology of Dansgaard et al. has been modified by stratigraphic correlation with dated records in North Atlantic deep-sea cores (Sancetta et al., 1973). Curve (d) is controlled by numerous ^{14}C dates.

ceding warm interval were destroyed in many places. Such vegetation records suggest that by the end of the Younger Dryas event, European climate had returned to about its present state.

The rise in sea level during the last 18,000 years indicated in Figure A.12(d) is generally ascribed to the melting of northern hemisphere continental ice sheets. Details of the sea-level curve, however, do not correspond to the chronology of deglaciation just described: while the continental ice sheets had essentially disappeared by about 7000 years ago, the worldwide stand of sea level has reached its maximum only during the last few thousand years or is still slowly rising

(Bloom, 1971). One possibility is that the West Antarctic ice sheet is unstable and has been disintegrating during the entire interval in question. Further research is clearly needed to settle this question, although it serves to illustrate the global interrelationships among the elements of the climatic system.

The Last 150,000 Years

In order to find an ancient counterpart to the warm, ice-free conditions of the past 10,000 years (the Holocene or present interglacial), it is necessary to go back some 125,000 years to an interval known as the Eemian interglacial (see Figure A.2). As shown by the proxy data of Figure A.13, the warmest part of this period lasted about 10,000

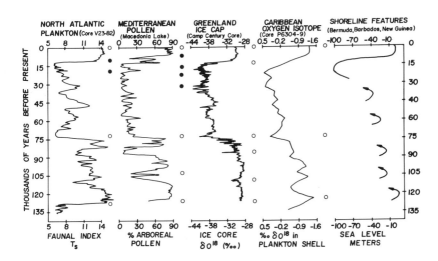

FIGURE A.13 Climatic records of the last 135,000 years. (a) A faunal index reflecting changes in foraminiferal plankton in a core west of Ireland. The index is an estimate of August sea-surface temperature in °C (Sancetta et al., 1973). (b) The percentage of tree pollen accumulated in a Macedonian lake. High values indicate warmer and somewhat dryer conditions (van der Hammen et al., 1971). (c) Oxygen isotope ratio expressed as $\delta^{18}O$ in an ice core at Camp Century, Greenland. This is interpreted as indicating changing air temperatures over the ice cap (Dansgaard et al., 1971). (d) Oxygen isotope ratio in skeletons of planktonic foraminifera in a Caribbean core, interpreted as changes in global ice volume. High negative values reflect the melting of ice containing isotopically light oxygen (Emiliani, 1968). (e) Generalized sea-level curve. Portion younger than 20,000 years is representative of a large number of sites (Bloom, 1971); see also Figure A.12. Older segments are from elevated coral reef tracts on Barbados, Bermuda, and New Guinea (Mesolella et al., 1969; Veeh and Chappell, 1970). Chronology of curves (a) to (d) controlled by ^{14}C dates (solid circles) and by stratigraphic correlation with dated horizons (open circles). Curve (e) is controlled by ^{14}C dates for the portion younger than 20,000 years and by uranium growth methods for the older segments.

years and was followed abruptly by a cold interval of substantial glacial growth lasting several thousand years. The interval between this post-Eemian event (c. 115,000 years ago) and the most recent glacial maximum 18,000 years ago was characterized by marked fluctuations superimposed on a generally declining temperature. An intense glacial event about 75,000 years ago is sometimes used to separate the interval into an older and generally nonglacial regime and a more recent glacial one.

A remarkable feature of the climatic record of the past 150,000 years is that both the present and the Eemian interglacials began with an abrupt termination of an intensely cold, fully glacial interval. Because these catastrophic episodes of deglaciation have left such a strong imprint on the climatic record, they have been named (in order of increasing age) termination I and termination II (see Figure A.14 and Broecker and van Donk, 1970).

The Last 1,000,000 Years

For at least the last 1,000,000 years the earth's climate has been characterized by an alternation of glacial and interglacial episodes, marked in the northern hemisphere by the waxing and waning of continental ice sheets and in both hemispheres by periods of rising and falling temperatures. How clearly these fluctuations are stamped on the various proxy data records is shown in Figure A.14. The most prominent features of the isotope curve shown here are seven terminations, marking a transition from full glacial to full interglacial conditions. All but one (termination III) of these changes are relatively rapid monotonic swings and provide an objective basis for defining a climatic "cycle" for at least the last 700,000 years. As shown in Figure A.14, these same fluctuations can be recognized in diverse and widely distributed records, including the chemical composition of Pacific sediments, fossil plankton in the Caribbean, and the soil types in central Europe. These "cycles," identified as A to E by Kukla (1970), are found in each of the records shown in Figure A.14 and may be grouped into a climatic "regime" covering the last 450,000 years (designated α). The earlier records (regime β) show higher-frequency fluctuations with less coherence among the various proxy climatic recorders.

The Last 100,000,000 Years

Although continuous and detailed records are lacking for these earlier times, at least the broad outline of this period of climatic history may

APPENDIX A 159

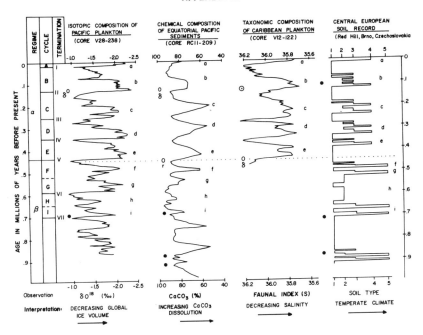

FIGURE A.14 Climatic records of the last 1,000,000 years. (a) Oxygen-isotope curve in Pacific deep-sea core V28-238, interpreted as reflecting global ice volume (Shackleton and Opdyke, 1973). The relatively rapid and high-amplitude fluctuations are taken to indicate sudden deglaciations and are designated as the terminations I to VII. (b) Calcium carbonate percentage in equatorial Pacific core RC11-209 (Hays et al., 1969). Low values are taken to indicate periods of rapid dissolution by bottom waters. (c) Faunal index reflecting changing composition of Caribbean foraminiferal plankton, calibrated as an estimate of sea-surface salinity in parts per thousand (Imbrie et al., 1973). Glacial periods are marked by the influx of plankton preferring higher-salinity waters (Prell, 1974). (d) Sequence of soil types accumulating at Brno, Czechoslovakia (Kukla, 1970). Type 1 is a wind-blown loess with a fossil fauna of cold-resistant snails or gley soils indicating extremely cold conditions; type 2 includes pellet sands and other hillwash deposits, partly interbedded with loess; type 3 includes brownearth and tschernosem soils; type 4 includes parabrownearth (lessivé) soils; type 5 are soils of temperate savannas, including brownlehms, rubified brownlehms, and rubified lessivés with large, hollow carbonate concretions. The duration of each soil type at this locality is plotted in proportion to the maximum thickness observed. Note that each record shown here reflects a climatic fluctuation or "cycle" averaging about 100,000 years. This is particularly true during the last 450,000 years (climatic regime α). Chronology of the curves is obtained by linear interpolation between indicated control points, shown by solid circles.

be discerned. From the climatic point of view, perhaps the most striking aspect of world geography at the beginning of this interval was the essentially meridional configuration of the continents and shallow ocean ridges, which must have prevented a circumpolar ocean current in either hemisphere. In the south this barrier was formed by South America and Antarctica (which lay in approximately their present latitudinal positions); by Australia (then a north-eastward extension of Antarctica); and by the narrow and relatively shallow ancestral Indian Ocean (Dietz and Holden, 1970). About 50,000,000 years ago the Antarctica–Australian passage began to open (Kennett et al., 1973), and as Australia moved northeastward, the Indian Ocean widened and deepened. Both paleontological and sedimentary evidence suggest that about 30,000,000 years ago the Antarctic circumpolar current system was first established. This must be considered a pivotal event in the climatic history of the past 100,000,000 years, and when the evidence of global plate movements is complete, it may well be possible to account for much of the secular climatic changes of this period as a response to the changing boundary conditions imposed by the distribution of land and ocean.

During the last part of the Mesozoic era (from 100,000,000 to 65,000,000 years ago) global climate was in general substantially warmer than it is today, and the polar regions were without ice caps. About 55,000,000 years ago numerous geologic records (Addicott, 1970; Flint, 1971) make it clear that global climate began a long cooling trend known as the Cenozoic climatic decline (see Figure A.15). Evidence from the marine record indicates that about 35,000,000 years ago Antarctic waters underwent a substantial cooling (Douglas and Savin, 1973; Shackleton and Kennett, 1974a, 1974b). There is direct evidence that ice reached the edge of the continent in the Ross Sea area some 25,000,000 years ago; and during the Oligocene epoch, roughly 35,000,000 to 25,000,000 years ago, global climate was generally quite cool (Moore, 1972). During early Miocene time (20,000,000 to 15,000,000 years ago) evidence from low and middle latitudes indicates a warmer climate, but isotopic evidence and faunal data indicate that this warming did not affect high southern latitudes.

About 10,000,000 years ago there is widespread evidence of further cooling, substantial growth of Antarctic ice (Shackleton and Kennett, 1974a, 1974b), and growth of mountain glaciers in the northern hemisphere (Denton et al., 1971). For general descriptive purposes the present glacial age may be defined as beginning at this time. Indirect evidence from marine sediments indicates that about 5,000,000 years ago the already substantial ice sheets on Antarctica underwent rapid growth and quickly attained essentially their present volume (Shackleton and Kennett, 1974a, 1974b). This evidence is generally consistent with direct records from the Antarctic continent, which show that between 7,000,000 and 10,000,000 years ago a large ice sheet existed in West Antarctica, and that by about 4,000,000 years ago the ice sheet in East Antarctica had developed to essentially its present volume

APPENDIX A 161

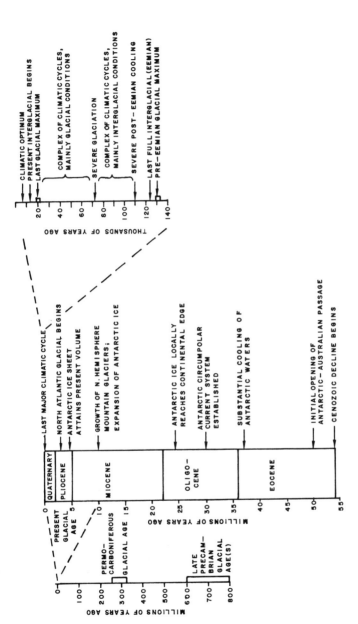

FIGURE A.15 History of glacial ages over the last 1,000,000,000 years. Intervals when extensive polar ice sheets occurred are indicated as glacial ages on the left. An outline of significant events in the Cenozoic climate decline is given in the middle, and the significant climate events during the last major glacial-interglacial cycle are given on the right.

(Denton et al., 1971; Mayewski, 1973). The present Antarctica ice mass is equivalent to about 59 m of sea level.

Although the behavior of the smaller, West Antarctic ice sheet is incompletely known, the available evidence indicates that it has undergone considerable fluctuation, and that its variations are roughly synchronous with the northern hemisphere glacial–interglacial cycle. This may be due to the fact that while the East Antarctic ice sheet is solidly grounded on the continent, much of the West Antarctic ice mass is grounded on islands or on the sea floor and could therefore be significantly influenced by sea-level variations due to glacial changes in the northern hemisphere. Such continental ice sheets first appeared in the northern hemisphere about 3,000,000 years ago, occupying lands adjacent to the North Atlantic Ocean (Berggren, 1972b), and during at least the last million years the ice cover on the Arctic Ocean was never less than it is today (Hunkins et al., 1971).

The Last 1,000,000,000 Years

Our knowledge of the climatic events over this time range consists principally of evidence of glaciations as preserved in the geological record. This may be seen in perspective with that for the more recent periods discussed above in Figure A.15. The present glacial age is seen to be at least the third time that the planet has suffered widespread continental glaciation. The Permo-Carboniferous glacial age (about 300,000,000 years ago) occurred at a time when the earth's land masses were joined in a single supercontinent (Pangaea). The area of this continent was distributed in roughly equal proportions between the hemispheres, with a concentration of land in the midlatitudes (Dietz and Holden, 1970). Glaciated portions of Pangaea included parts of what are now South America, Africa, India, Australia, and Antarctica. One or more earlier glacial ages are known from late Precambrian times (about 600,000,000 years ago), from the indications of glaciation in deposits now widely scattered over the globe, including Greenland, Scandinavia, central Africa, Australia, and eastern Asia (Holmes, 1965).

Although other glacial ages may have occurred besides those recognized in Figure A.15, none has left such a clear and widespread imprint on the geological record (Steiner and Grillmair, 1973). While the evidence is far from complete, it may be that each of the earth's major glacial ages—including the present one—resulted from crustal movements that permitted the development of sharp thermal gradients over a continental land mass that includes a pole of rotation. To establish this or other hypotheses of long-period climatic changes, however, will require the assembly of a much more complete geological record and the performance of appropriate climate modeling experiments.

GEOGRAPHIC PATTERNS OF CLIMATIC CHANGE

While the chronology of certain features of climatic change may be revealed by the analysis of instrumental and paleoclimatic data at individual sites, the geographic pattern of these changes is an equally important characteristic. From what we know of the behavior of the (present) atmosphere, it would be remarkable if there were not a definite spatial structure to the variations on all climatic times scales. The search for these patterns requires synoptic data for the various climatic elements, and this is presently available only from the records of modern observations and from a few marine proxy sources.

Structure Revealed by Observational Data

The task of describing the spatial and temporal structure of climatic variations from the observations of the instrumental era is far from complete. Most studies have therefore focused primarily on local or regional climatic changes. Lamb and Johnson (1961, 1966) have made comprehensive analyses of intrahemispheric and interhemispheric climatic indices, and the statistical structure of these circulation variations has been studied by Willett (1967), Wagner (1971), Iudin (1967), Brier (1968), and Kutzbach (1970). Such analyses, especially of hemispheric pressure data, reveal that the year-to-year and decade-to-decade variations have a spatial structure that may be associated with amplitude and phase changes of the long planetary waves in the atmosphere.

The essentially two-dimensional character of climate is masked in studies of zonally averaged parameters, although these may be useful for other purposes. An example of the importance of both zonal and nonzonal spatial variability of the atmospheric circulation is provided by the first eigenvector pattern (or empirical orthogonal function) of hemispheric pressure for January, shown in Figure A.16, as well as by the patterns of pressure, temperature, and rainfall variability shown in Figures A.17, A.18, and A.19. These data suggest an association between the changes in the monthly average intensity and position of the Aleutian and Icelandic lows. For example, during the first two decades of this century there has been a tendency for decreased intensity and westward extension of the Aleutian low, coupled with an

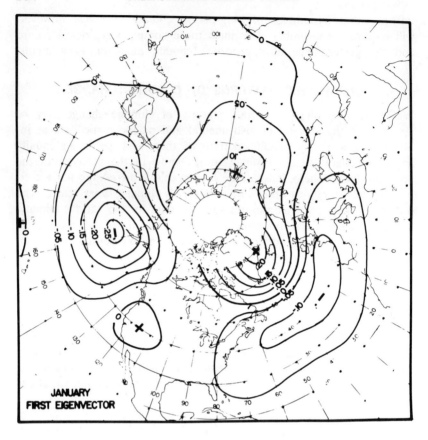

FIGURE A.16 The first eigenvector of northern hemisphere sea-level pressure, based on the individual mean January maps for 1899–1969 (Kutzbach, 1970). This spatial function accounts for 22 percent of the total inter-January variance.

increased intensity and northeastward shift of the Icelandic low. Lamb (1966) and Namias (1970) have described important regional changes in temperature and precipitation associated with these circulation changes. The opposite tendency has prevailed since the mid-1950's, and Lamb (1966), Winstanley (1973), and Bryson (1974) have described the possible relationships between the changing midlatitude circulation patterns of the 1960's, the equatorial shift of the subtropical highs, and the increasing frequency of droughts along the southern fringes of the monsoon lands of the northern hemisphere (see Figure A.8). These changes appear to reflect an equatorward extension of the westerly wave regime and a contraction of the Hadley circulation,

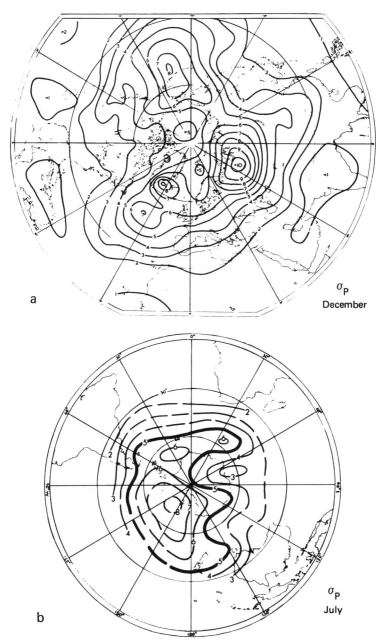

FIGURE A.17 Standard deviation (mbar) of monthly mean pressure at sea level, 1951–1966 (Lamb, 1972). (a) December, northern hemisphere; (b) July, southern hemisphere.

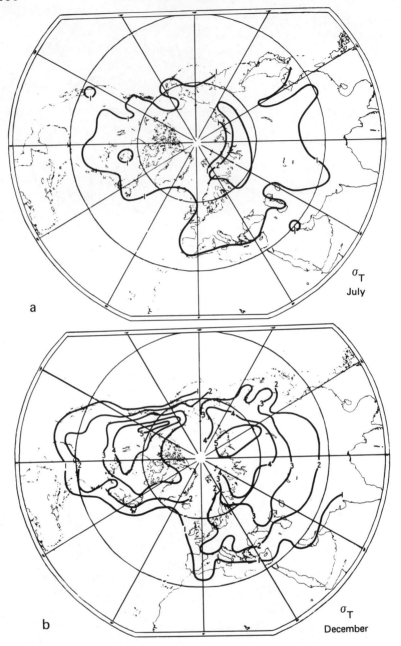

FIGURE A.18 Standard deviation (°C) of monthly mean surface air temperature in the northern hemisphere, 1900–1950 (Lamb, 1972). (a) July; (b) December.

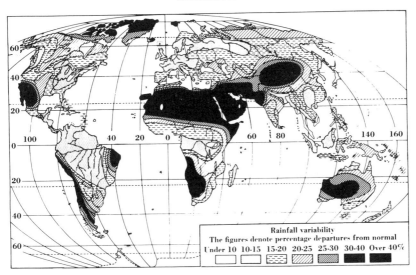

FIGURE A.19 The variability of mean annual rainfall for the world (adapted from Petterssen, 1969).

although much further analysis is clearly required to confirm such a conjecture.

Interhemispheric relationships of climatic indices have been (and remain) less amenable to study because of the general lack of observations from the southern hemisphere. Observations are sufficient, however, to show that the circulation in the southern hemisphere is somewhat stronger and steadier than that in the northern hemisphere. Whether this results in the southern hemisphere circulation leading that in the northern hemisphere, or whether variable features in the equatorial circulation influence both hemispheres similarly, is not presently known (Bjerknes, 1969b; Fletcher, 1969; Lamb, 1969; Namias, 1972a, 1972b). It is likely that interhemispheric relationships of one sort or another are important for the understanding of climatic variations, and that our ability to describe them will require the availability of much more comprehensive data than now exist from the southern hemisphere, the equatorial region, and the oceanic and polar regions of the northern hemisphere.

The present accumulation of upper-air data, especially in the northern hemisphere since the early 1950's, however, has permitted a beginning of the study of the three-dimensional spatial and temporal variability of the general circulation. A foundation of basic statistics is provided by calculations of the means and variances of standard meteorological variables (see, for example, Crutcher and Meserve, 1970; Taljaard *et al.*, 1969) and by atlases of energy budgets (Budyko, 1963). The covariance structure of circulation patterns at 700 mbar in the northern hemisphere is treated by O'Connor (1969), and other aspects of the tropospheric circulation have been considered by Gommel (1963) and Wahl (1972). The most comprehensive analysis of atmospheric circulation statistics, however, is that based on the period 1958–1963 as undertaken by Oort and Rasmusson (1971). While this work documents the monthly, seasonal, and annual variations of many features of the observed general circulation (in the northern hemisphere), it does not directly address many of the variables of primary climatic interest. Using the same data set, however, Starr and Oort (1973) have reported an unmistakable downward trend of the mean air temperature in the northern hemisphere of 0.6°C over the five-year interval shown in Figure A.20. Diagnostic studies of this type represent great investments of time and effort but are essential steps toward the monitoring of climate and an assessment of the mechanisms of climatic variation.

A complete description of climatic changes from instrumental records must also include studies of the momentum and energy budgets of the atmosphere and oceans and their variability with time over many years and decades. While this must remain largely a task for the future, several efforts have established the existence of significant interannual variations in the atmosphere. Krueger *et al.* (1965) have discussed the

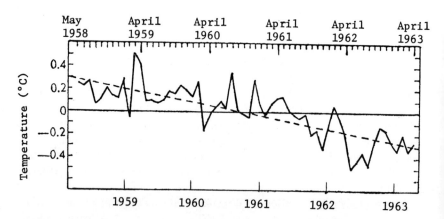

FIGURE A.20 Monthly mean mass-averaged values of the northern hemisphere temperature for the period May 1958 to April 1963 (Starr and Oort, 1973). The consecutive monthly averages are plotted on the scale marked at the top; the bottom scale shows the beginning of each calendar year.

interannual variations of available potential energy, and Kung and Soong (1969) have described the fluctuations of the atmospheric kinetic energy budget. As noted previously, the interannual variations of poleward angular momentum and energy fluxes has been studied comprehensively by Oort and Rasmusson (1971). A measure of this variability is shown in Figure A.21 and amounts to about ±30 percent of the mean transports.

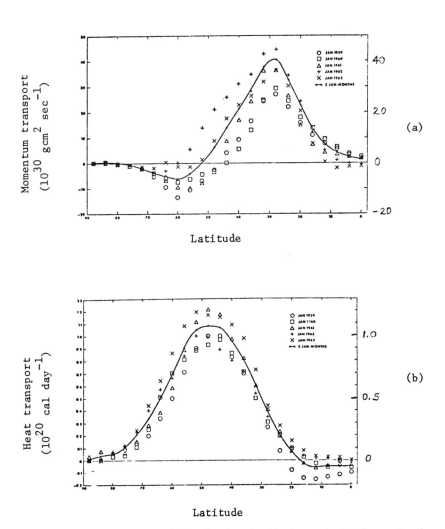

FIGURE A.21 Interannual variability of poleward eddy transports for the month of January as shown by five years data (Oort and Rasmusson, 1971). The solid curve is the 5-year mean for 1959–1963. (a) Angular momentum transport; (b) sensible heat transport.

The unique global potential of satellite-based measurements has been exploited by Vonder Haar and Suomi (1971), who have summarized the satellite measurements of planetary albedo and of the planetary radiation budget for the five years 1962 to 1966. They found large interannual variations in the zonally averaged equator-to-pole gradient of the net radiation as shown in Figure A.22. This forcing function can now be monitored routinely by meteorological satellites and opens the door to more detailed studies of atmospheric energetics than heretofore possible (Winston, 1969). Vonder Haar and Oort (1973) have combined satellite measurements of the earth's radiation budget with atmospheric energy transport calculations to produce a new estimate of the poleward energy transport by the northern hemisphere

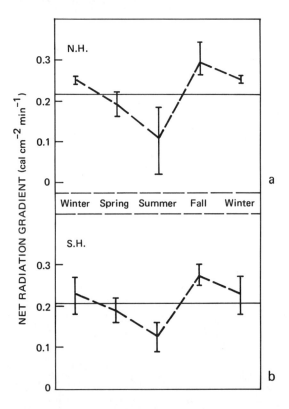

FIGURE A.22 Seasonal (dashed lines) and interannual variation (vertical bars) of the equator-to-pole gradient of net radiation as measured from satellites (Vonder Haar and Suomi, 1971). (a) Northern hemisphere; (b) southern hemisphere.

APPENDIX A 171

oceans. They find that the oceanic heat transport averages about 40 percent of the total in the 0–70 °N latitude band and accounts for more than half at many latitudes. Another example of the use of satellite-derived measurements of climatic indices is given by Kukla and Kukla (1974). Their measurements of the interannual changes in the area of snow and ice cover in the northern hemisphere are shown in Figure A.23 and reveal year-to-year fluctuations of the order of a few percent. Note, however, the relatively large change during 1971 and the subsequent maintenance of extensive snow and ice coverage and an associated increase of the reflected solar radiation.

Time variations of the *surface* energy budget on a global scale are

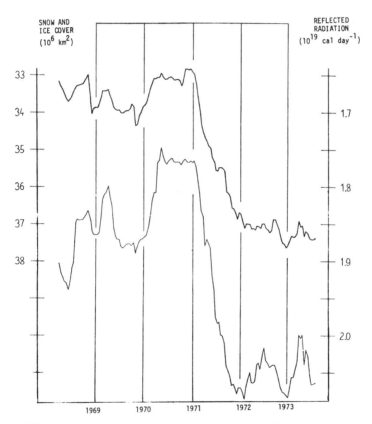

FIGURE A.23 Twelve-month running means of snow and ice cover in the northern hemisphere (upper curve) and the estimated reflected solar radiation disregarding variations of cloudiness (lower curve), as reported by Kukla and Kukla (1974). The averages are plotted on terminal dates, with the years marking January 1.

not available from direct observations and must be inferred from the conventional measurements of temperature, humidity, cloudiness, wind, and radiation. Fletcher (1969) has drawn attention to the variations in the energy budget of polar regions as a function of variable sea-ice conditions, while Sawyer (1964) has noted the possible role of fluctuations in the surface energy budget as a cause of interannual variations of the general circulation itself.

A number of observational studies of large-scale interaction between the ocean and the atmosphere have illustrated the complexity and importance of this mechanism; see, for example, Weyl (1968) and Lamb and Ratcliffe (1972). Bjerknes (1969b) has considered the response of the North Pacific westerlies to anomalies of equatorial sea-surface temperature and variations in the Hadley circulation, while Namias (1969, 1972b) has described positive feedback relationships between large-scale patterns of ocean-surface temperature in midlatitudes and the circulation of the overlying atmosphere. Such modes of atmosphere–ocean coupling may be important parts of climatic fluctuations and must be given further study.

In summary, we may say that observational data at the earth's surface show that during the period 1900 to 1940 the northern hemisphere as a whole warmed, although some areas (mainly the Atlantic sector of the Arctic and northern Siberia) warmed far more than the global average, some areas became colder, and others showed little measurable change (Mitchell, 1963). In the time since 1940, an overall cooling has occurred but is again characterized by a geographical structure; cooling since 1958 has occurred in the subtropical arid regions and in the Arctic (Starr and Oort, 1973). There is also some evidence that the northern hemisphere oceans are cooling (Namias, 1972b), although the oceanic data base necessary to confirm this has not yet been assembled.

Structure Revealed by Paleoclimatography

Most of the work done to date on climatic change beyond the time frame encompassed by meteorological observations represents a study of time series taken at specific sites. This lack of synoptic data on the longer-range climatic changes is a serious handicap to the portrayal and understanding of the mechanisms involved. In order to underscore these points, and to encourage further research, we present here examples of the few proxy data that have been assembled to reveal a spatial structure of climatic change.

Distribution of Ice Sheets

The continental margins of the northern hemisphere ice sheets at their maximum extension during the last million years are clearly marked by the debris deposits in terminal moraines, while the extent of sea ice is recorded by features preserved in marine sediments. Figure A.24

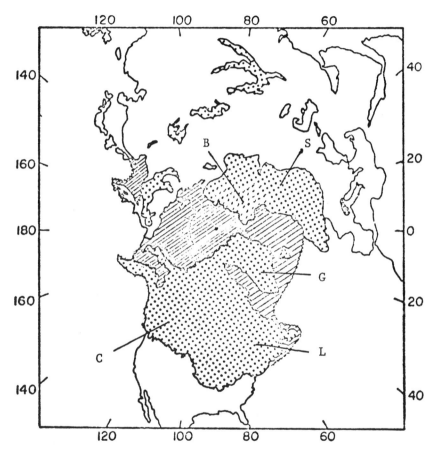

FIGURE A.24 Maximum extent of northern hemisphere ice cover during the present glacial age (modified after Flint, 1971). Continental ice sheets, indicated by the dotted area, are B, Barents Sea; S, Scandinavian; G, Greenland; L, Laurentide; C, Cordilleran. Sea ice is indicated by the cross-hatched pattern. The boundary mapped is the southernmost extent of the ice margin that occurred in any sector during the last million years. The last glacial maximum, about 18,000 years ago, occupied about 90 percent of the area shown here (see also Table A.2).

shows the distribution of maximum ice cover, and Table A.2 gives statistics of the areas of the individual continental ice sheets. In North America the ice extended as far south as 40 °N and spanned the entire width of the continent, while in Europe the ice sheet extended only to about 50 °N. Note that large regions in eastern Siberia were unglaciated.

Sea-Surface Temperature Patterns

The north–south migration of polar waters in the North Atlantic in response to major cycles of glaciation is shown in Figure A.25. During glacial maxima these waters were found as far south as 40 °N, well beyond the present extent of polar waters. A synoptic analysis of the ocean surface temperatures of 18,000 years ago (at about the time of the last glacial maximum) is shown in Figure A.26. These temperature estimates have been derived by multivariate statistical techniques applied to planktonic organisms as preserved in about 100 deep-sea cores in scattered locations across the North Atlantic (McIntyre et al., 1974). The most striking feature of this glacial-age map is the extensive southward displacement of the 10 to 14°C water, while the warmer water was found in nearly its present position. In parts of the Sargasso Sea the glacial-age ocean was, if anything, slightly warmer than it is today.

Because the atmosphere receives much of its heat from the sea, such estimates of sea-surface temperature are likely to be important in developing a satisfactory reconstruction of past climates, and it is therefore important to consider their reliability. Berger (1971), for example, has suggested that carbonate dissolution on the sea bed may distort the taxonomic composition of the fossil fauna on which such paleotemperature estimates are based. Kipp (1974), on the other hand, shows that when the statistical transfer functions are calibrated on materials that incorporate the dissolution effects, an unbiased estimate of such parameters as the sea-surface temperature can be obtained. The temperature reconstruction in Figure A.26(b) is based on the statistics of the foraminiferal fauna distribution and encompasses 91 percent of the variance of the data (McIntyre, 1974). The 80 percent confidence interval of each of the cores is ± 1.8°C (Kipp, 1974). Shackleton and Opdyke (1973), using a revised isotopic method based on the difference between ^{18}O values in benthic and planktonic species, have provided an independent confirmation of the sea-surface temperature estimates of Imbrie et al. (1973) for a portion of the glacial-age Caribbean.

Other reconstructions of paleo-ocean surface temperatures have been based on data from radiolaria, coccoliths, and foraminifera; and al-

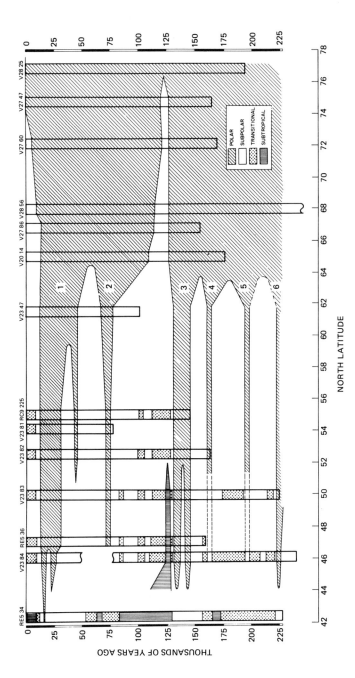

FIGURE A.25 Evidence of north-south migration of polar water during the past 225,000 years as shown by 14 deep-sea cores arranged along a transect in the eastern North Atlantic (Kellogg, 1974; McIntyre et al., 1972). The boundary between the polar fossil assemblages (diagonal ruling) and the subpolar assemblages (open pattern) reflects the position of the polar front, which was some 20° farther south during the last glacial maximum about 18,000 years ago. The numbers 1 to 6 indicate calcium carbonate sediment minima.

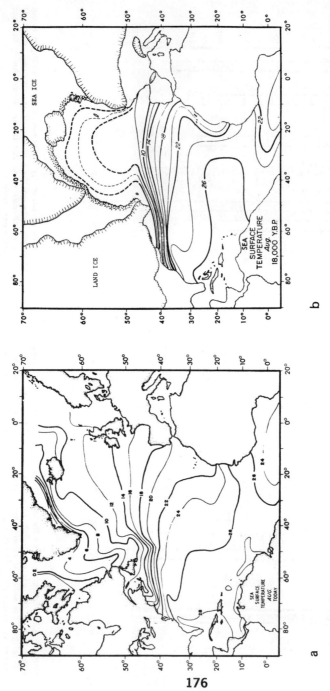

FIGURE A.26 Sea-surface temperatures in the North Atlantic. (a) Summer temperatures in the North Atlantic. (a) Summer temperatures today in °C (from McIntyre et al., 1974). (b) Summer temperatures 18,000 years ago in °C (from McIntyre et al., 1974) as calibrated by a faunal index reflecting variations in the composition of foraminiferal plankton in sediment taken from about 100 deep-sea cores. Land-ice margins are taken from Flint (1971), with sea-level assumed to be 100 m lower than at present. The sea-ice margins are identified from sediment characteristics. Note that there has been substantial cooling in the northern waters, with little or no change in the subtropics.

though some discrepancies are revealed where independent data are available, the derived ocean temperatures show considerable spatial coherence (McIntyre, 1974). Such estimates of past sea-surface temperature will prove useful in climatic simulations with numerical general circulation models (see Appendix B).

Patterns of Vegetation Change

Figure A.27 illustrates the use of fossil pollen data to record the changes in vegetation accompanying the deglaciation of eastern North America during the interval 11,000 to 9000 years ago. At the beginning of this time, pine species occupied sites in the southeastern Appalachians, but as the ice retreated, the pine moved farther north and west to colonize newly uncovered areas. A relatively complete chronology of the retreat of the Laurentide ice sheet itself is given by radiocarbon dating (Bryson et al., 1969).

Patterns of Aridity

For only four desert areas in the world do we have enough information to plot aridity as a function of time, and even in these areas the record extends back only a few tens of thousands of years. As shown in Figure A.28, the data suggest a degree of synchroneity between the two African regions and the Great Basin, while the records from the Middle East are quite different. None of the data from closed-basin lakes show significant correlation with the glacial record, and we are clearly a long way from understanding the response of arid regions to glacial cycles. More generally, insufficient research has been devoted to the role of desert regions in the processes responsible for the climate of the earth.

Patterns of Tree-Ring Growth

Changes of thickness of the growth rings added by trees each year reflect environmental change in a complex way. By appropriate calibration, such data may be made to furnish significant climatic information for the past several hundred to several thousand years. Studies of many tree-ring series over a wide geographic area can, moreover, provide accurately dated synoptic evidence of regional climatic patterns (Fritts, 1965).

Fritts et al. (1971) have demonstrated the feasibility of reconstructing the anomalies of sea-level pressure and temperature from the spatial patterns of tree growth over western North America. Examples

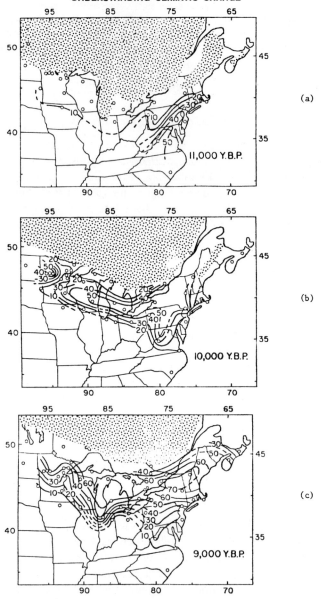

FIGURE A.27 The distribution of pine pollen at selected times during the deglaciation of eastern North America (Bernabo et al., 1974). Contours are lines of pollen frequency, expressed as a percent of total pollen. Control points representing radiocarbon dated cores are indicated by the open circles. The approximate margins of the Laurentide ice sheet are indicated by the stippled pattern (after Bryson et al., 1969).

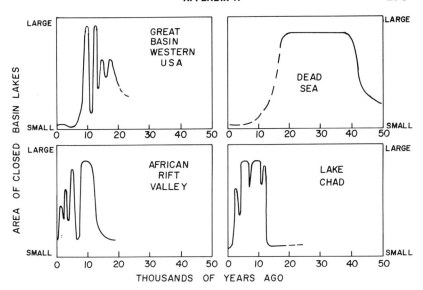

FIGURE A.28 Variations in the size of closed-basin lakes, as indicated by the degree of aridity found from radiocarbon dating of shorelines and bottom sediments. Higher rainfall and lower evaporation may be inferred at the times of larger water surfaces. (a) From Broecker and Kaufman (1965); (b) from Kaufman (1971); (c) from Butzer et al. (1972); (d) from Servant et al. (1969).

of such synoptic maps based on average decadal growth are given in Figure A.29. Although such reconstructions show considerable variation in the year-to-year climatic states, the inferred variations in the intensity of Icelandic and Aleutian lows, for example, are similar to those described in the modern record (Kutzbach, 1970). The development of an expanded network of tree-ring sites could significantly broaden our knowledge of the patterns of climatic fluctuations over the past several centuries.

SUMMARY OF THE CLIMATIC RECORD

In this survey of past climates, the characteristic time and spatial structures of climatic variations have been discussed as though there were sufficient data to document large regions of the globe. This is true only for the more recent parts of the instrumental period, as there are large gaps in the presently available historical and proxy climatic records. With these limitations in mind, it is nevertheless useful to summarize the general characteristics of the climatic record:

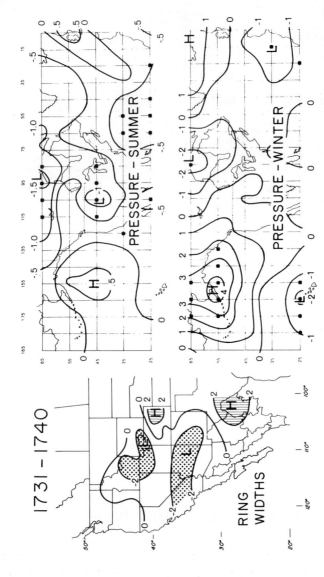

FIGURE A.29 Anomalies of ten-year average tree-ring widths (left), based on 49 tree stands distributed over western North America and the statistically inferred patterns of seasonal surface pressure anomalies (in mbar, right). The calibration is obtained from an analysis of modern recorded pressure data and then applied to earlier ring-width data to reconstruct climatic conditions. Squares indicate statistically significant departures from normal in the reconstructed pressures. In general, higher-than-normal growth indicates a relatively cool and moist local climate (from Fritts et al., 1971).

1. The last postglacial thermal maximum was reached about 6000 years ago, and climates since then have undergone a gradual cooling. This trend has been interrupted by three shorter periods of more marked cooling, simliar to the so-called Little Ice Age of A.D. 1430–1850, each followed by a temperature recovery. The well-documented warming trend of global climate beginning in the 1880's and continuing until the 1940's is a continuation of the warming trend that terminated the Little Ice Age. Since the 1940's, mean temperatures have declined and are now nearly halfway back to the 1880 levels.

2. Climatic changes during the past 20,000 years are as severe as any that occurred during the past million years. At the last glacial maximum, extensive areas of the northern hemisphere were covered with continental ice sheets, sea level dropped about 85 m, and sea-surface temperatures in the North Atlantic fell by as much as 10°C. At northern midlatitude sites not far from the glacial margins (locations now occupied by major cities and extensive agricultural activity), air temperatures fell markedly, drastic changes occurred in the precipitation patterns, and wholesale migrations of animal and plant communities took place.

3. The present interglacial interval—which has now lasted for about 10,000 years—represents a climatic regime that is relatively rare during the past million years, most of which has been occupied by colder, glacial regimes. Only during about 8 percent of the past 700,000 years has the earth experienced climates as warm as or warmer than the present.

4. The penultimate interglacial age began about 125,000 years ago and lasted for approximately 10,000 years. Similar interglacial ages—each lasting $10,000 \pm 2000$ years and each followed by a glacial maximum—have occurred on the average every 100,000 years during at least the past half million years. During this period fluctuations of the northern hemisphere ice sheets caused sea-level variations of the order of 100 m. In contrast, the East Antarctic ice sheet has apparently varied little since reaching its present size about 5 million years ago, while the West Antarctic ice sheet appears to have been disintegrating for many thousands of years.

5. About 65 million years ago global climates were substantially warmer than today, and subsequent changes may be viewed as part of a very long-period cooling trend. For even earlier times, the proxy climatic evidence becomes increasingly fragmentary. The best documented records suggest two previous extensive glaciations, occurring about 300 million and 600 million years ago.

FUTURE CLIMATE: SOME INFERENCES FROM PAST BEHAVIOR

The overall picture of past climatic changes described in this survey suggests the existence of a hierarchy of fluctuations that stand out above the "white noise" or random fluctuations presumed to exist on all time scales. In addition to the dominant period of about 100,000 years, there are apparent quasi-periodic fluctuations with time scales of about 2500 years and shorter-period fluctuations on the order of 100–400 years. Each of these explains progressively less of the total variance but may nevertheless be climatically significant. No periodic component of climatic change on the order of decades has yet been clearly established, although significant excursions of climate are observed to occur in anomalous groups of years.

In view of the limited resolving power of most climatic indicators, especially those for the relatively remote geological past, it is difficult to establish whether the apparent fluctuations are quasi-periodic or whether they more nearly represent what are basically random Markovian "red-noise" variations. In the case of the longer-period variations (of 100,000-year and 20,000-year periods), there is circumstantial evidence to suggest that these may have been induced in some manner by the secular variations of the earth's orbital elements, which are known to alter the seasonal and latitudinal distribution of solar radiation received at the top of the atmosphere. In other cases, the observed variations have yet to be convincingly related to any external climatic control. The mere existence of such variations does not necessarily mean that changes in the external boundary conditions are involved, however. The internal dynamics of the climatic system itself may well be the origin of some of these features. Whether forced or not, climatic behavior of this type deserves careful study, as the conclusions reached bear directly upon the problem of inferring the future climate.

The prediction of climate is clearly an enormously complex problem. Although we have no useful skill in predicting weather beyond a few weeks into the future, we have a compelling need to predict the climate for years, decades, and even centuries ahead. Not only do we have to take into account the complex year-to-year changes possibly induced by the internal dynamics of the climatic system, and the likely continuation of the (yet unexplained) quasi-periodic and episodic fluctuations of the last few thousand years discussed above, but also the changes induced by possibly even less predictable factors such as the aerosols added to the atmosphere by volcanic eruptions and by man himself (Mitchell, 1973a, 1973b). These questions lie at the heart of the problem of climatic variation and are given consideration elsewhere in this report.

In the face of these uncertainties, any projection of the future climate carries a great risk. Nevertheless, we may speculate about the possible course of global climate in the decades and centuries immediately ahead by making certain assumptions about the character of the major fluctuations noted in the climatic record. In the following paragraphs we attempt to draw together these considerations into an overall assessment of the probable direction and magnitude of present-day climatic change, taking into account the risk of a major future change associated with the seemingly inevitable onset of the next glacial period.

Potential Contribution of Sinusoidal Fluctuations of Various Time Scales to the Rate of Change of Present-Day Climate

Estimates of the amplitudes of all the principal climatic fluctuations identified in this report are listed in Table A.3 (where they have been made consistent with the data presented in Figure A.2 and are expressed in terms of the total range of temperature between maxima and minima). On the assumption that all of these fluctuations can be approximated by quasi-periodic sine waves, the ratio of the amplitude (A) to the period (P) of each fluctuation becomes proportional to the *maximum* contribution of that fluctuation to the rate of change of climate. By considering also the phase of each fluctuation, as inferred from the paleoclimatic record, the contribution of each fluctuation to the present-day rate of change can be estimated (see Table A.3).

Estimation of the phase of each sinusoidal fluctuation (indicated by the estimated dates of the last temperature maximum in Table A.3) permits an assessment of the sign and magnitude of the contribution of each fluctuation to the total rate of change of globally average temperature in the present epoch. The sum of these individual contributions ($-0.015\,°C/yr$) agrees reasonably well with the observed rate of change of $-0.01\,°C/yr$ during the past two decades, as determined from analyses of surface climatological data by Reitan (1971) and by Budyko (1969). It should be noted that this trend is dominated by the shortest fluctuations, and especially by the fluctuations of the order of 100 years (see Figure A.6).

The estimated maximum rate of change associated with all time scales of climatic fluctuation shown in Figure A.2 is plotted as a continuous function of wavelength in Figure A.30. The family of curves also shown in this figure indicates the relationship between maximum

TABLE A.3 Estimated Characteristics of Principal Climatic Fluctuations (Quasi-periodic Model)

Approximate Wavelength (yr) P	Estimated Mean Double Amplitude (Temperature in °C), 2A	Estimated Date of Last Temperature Maximum (yr Before Present)	Present Temperature Level	Rate of Temperature Change [a] (°C/yr)	
				Maximum ($2\pi A/P$)	Circum. Decade of 1970's
100,000	8	10,000	Very high	±0.00025	−0.00015
20,000	3	8,000	High	±0.00045	−0.0003
2,500	2	1,750	Average	±0.0025	+0.0024
c. 200	0.5	75	High	±0.0075	−0.0053
c. 100	0.5	35	Average	±0.0150	−0.0121
			SUMS	±0.0257	−0.0154 °C/yr

[a] Assumes sinusoidal shape of fluctuations.

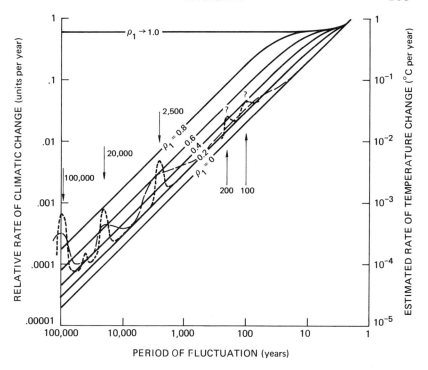

FIGURE A.30 Relative (maximum) rate of change of climate contributed by climatic fluctuations, as a function of characteristic wavelength. The family of parallel curves shows the expected relationship in Markovian "red noise" as characterized by the serial correlation coefficient at a lag of one year. The dashed line is a conservative estimate of actual climate as inferred from the data of Figure A.2 and from additional data on the shorter-period fluctuations from Kutzbach and Bryson (1974). The dotted curve shows the modifications to be expected if the principal fluctuations identified in Table A.3 were actually quasi-periodic.

rate of change and wavelength in Markovian "red noise," for various degrees of "redness" characterized by the value of the serial correlation coefficient at a time lag of one year (Gilman *et al.*, 1963). By comparison with these curves, it is suggested that the observed shorter-period climatic fluctuations (i.e., fluctuations of the order of 100 to 200 years) are not clearly distinguishable from random fluctuations, whereas the longer-period fluctuations (especially those with periods of 20,000 years or more) may be appreciably larger in amplitude than would be expected in random noise. The contributions of the longer-period fluctuations to present-day climatic change are seen nonetheless to be relatively small. Should the longer-period fluctuations be nonsinusoidal (or episodic) in form, rates of change perhaps ten times larger than the magnitudes shown in Figure A.30 could be possible. Even such rates, however, would contribute little over and above the normal interannual variability of present-day global climate, and the cumulative change of climate associated with the longer-period fluctuations would remain relatively small until several centuries had elapsed.

Despite its simplistic view of climatic change, this exercise is an instructive one in that it demonstrates how difficult it would be for longer-period sinusoidal fluctuations to contribute substantially to the changes of climate taking place in the twentieth century. If the longer-period fluctuations are those that primarily determine the course of the glacial–interglacial succession of global climate, it would seem that the transition to the next glacial period—even if it has already commenced—will require many centuries to accumulate to a drastic shift from present climatic conditions.

In assessing such projections, however, we must keep in mind that our ability to anticipate the locally important synoptic pattern of climatic variations is limited. The work of Mitchell (1963), for example, has shown that while the northern hemisphere average air temperatures rose only about 0.2°C during the period 1900 to 1940, there were many areas that deviated markedly from this hemispheric average trend. Parts of the eastern United States, for example, exhibited a 1.0°C rise in average temperature (5 times the hemispheric average), parts of Scandinavia and Mexico showed temperature increases of 2.0°C (10 times the hemispheric average), while in Spitsbergen the warming was 5°C (25 times the hemispheric average). The corresponding data on other climatic elements are sparse but may be expected to exhibit comparable or even greater spatial variance.

Likelihood of a Major Deterioration of Global Climate in the Years Ahead

As noted above, the longer-period climatic fluctuations seem to be associated with larger amplitudes of change than those consistent with Markovian "red-noise" behavior. The same cannot be said, however, of the shorter-period fluctuations. For the moment let us suppose that *all* the fluctuations described in this report are actually random fluctuations, in the sense that transitions between successive maxima and minima may occur at random (Poisson-distributed) intervals of time rather than at more or less regular intervals. The probability that one or more transitions of a fluctuation will occur in an arbitrarily specified length of time may then be calculated from the negative binomial distribution. Following this approach, we can assess the risk of encountering a change of climate in the years ahead as rapid as the maximum rate of change otherwise associated with sinusoidal climatic fluctuations on each of the characteristic time scales noted above. Such a measure of risk, for time intervals between 1 year and 1000 years into the future, can be inferred by interpolation between the curves of transition probability in Figure A.31. The proper interpretation of this figure will be apparent from the following examples:

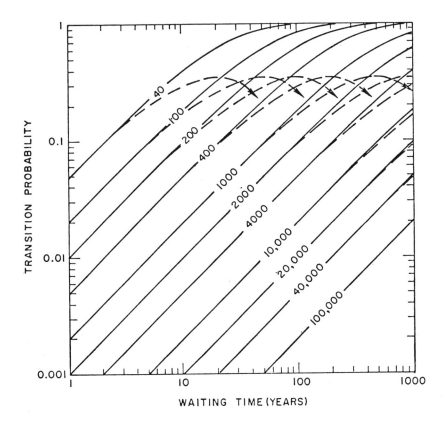

FIGURE A.31 Probability of onset of climatic transitions analogous to the changes between maxima and minima in climatic fluctuations of arbitrarily selected characteristic wavelengths (interior numbers, in years), as a function of elapsed time after present. Dashed curves denote probability of one transition; solid curves denote that of one or more transitions. Based on the assumption that intervals between transitions are strictly random (Poisson distributed).

1. The curve labeled 100,000 in the figure indicates the probability of a major transition of climate (in either direction) that is normally associated with climatic fluctuations on the time scale of 100,000 years (a change of global average temperature of up to perhaps 8°C in a total time interval of 50,000 years or less). The curve indicates that if successive transitions of this kind recur at *random* time intervals as assumed here, the onset (or termination) of such a transition will occur in the next 100 years with a probability of about 0.002 and in the next 1000 years with a probability of about 0.02.

2. The dashed curve labeled 100 in the figure indicates the probability of *one* transition of climate (in either direction) that is normally associated with climatic fluctuations on the time scale of 100 years (a change of up to perhaps 0.5°C in a total time interval of about 50 years or less). Such a transition is indicated to have a probability of about 0.02 of occurring in the next year, a probability of about 0.16 of occurring in the next 10 years, and a probability of about 0.35 of occurring in the next 50 years. The solid line labeled 100 in the figure indicates the probability of *one or more* transitions of the same kind, which rises from about 0.2 in the next 10 years to about 0.8 in the next 100 years. If it can be assumed that the typical duration of such a transition (when it occurs) is not less than four or five decades, and that only one such transition can occur at the same time, then the dashed curve would be the appropriate guide for estimating such probabilities in the next few decades. Otherwise, the solid curve would be a more appropriate guide.

When Figures A.30 and A.31 are considered together, it is suggested that whether climatic fluctuations are or are not quasi-periodic, those that are most relevant to the course of global climate in the years and decades immediately ahead are the shorter-period (historical) fluctuations and not the longer-period (glacial) fluctuations. Even if the phase of the longer-period changes is such as to contribute to a cooling of present-day climate, the contribution of such fluctuations to the rate of change of present-day climate would seem to be swamped by the much larger contributions of the shorter-period (if more ephemeral) historical fluctuations. We must remember, however, that this analysis assumes a simple model of climatic change in which climatic fluctuations of various periods are independent and therefore additive. The paleoclimatic record presented here does not preclude the possibility that relatively sudden climatic changes could arise through interactions between fluctuations of different periods.

One may still ask the question: When will the present interglacial end? Few paleoclimatologists would dispute that the prominent warm periods (or interglacials) that have followed each of the terminations of the major glaciations have had durations of $10,000 \pm 2000$ years. In each case, a period of considerably colder climate has followed immediately after the interglacial interval. Since about 10,000 years has elapsed since the onset of the present period of prominent warmth, the question naturally arises as to whether we are indeed on the brink of a period of colder climate. Kukla and Matthews (1972) have already called attention to such a possibility. There seems little doubt that the present period of unusual warmth *will* eventually give way to a time of colder climate, but there is no consensus with regard to either the magnitude or rapidity of the transition. The onset of this climatic decline could be several thousand years in the future, although there is a finite probability that a serious worldwide cooling could befall the earth within the next hundred years.

What is the nature of the climatic changes accompanying the end of a period of interglacial warmth? From studies of sediments and soils, Kukla finds that major changes in vegetation occurred at the end of the previous interglacial (Figure A.14). The deciduous forests that covered areas during the major glaciations were replaced by sparse shrubs, and dust blew freely about. The climate was considerably more continental than at present, and the agricultural productivity would have been marginal at best. The stratification of fossil pollen deposits in eastern Macedonia (Figure A.13) also clearly shows a marked change in vegetative cover between interglacial warmth and the following cold periods. The oak–pine forest that existed in the area gave way to a steppe shrub, and grass was the dominant plant cover. Other evidence from deep-sea cores reveals a substantial change in the surface water temperature in the North Atlantic between interglacial and glacial periods (Figure A.13), and the marine sediment data show that the magnitude of the characteristically abrupt glacial cooling was approximately half the total glacial to interglacial change itself.

The question remains unresolved. If the end of the interglacial is episodic in character, we are moving toward a rather sudden climatic change of unknown timing, although as each 100 years passes, we have perhaps a 5 percent greater chance of encountering its onset. If, on the other hand, these changes are more sinusoidal in character, then the climate should decline gradually over a period of thousands of years. These are the limits that we can presently place on the nature of this transition from the evidence contained in the paleoclimatic record.

These climatic projections, however, could be replaced with quite different future climatic scenarios due to man's inadvertent interference with the otherwise natural variation (Mitchell, 1973a). This aspect of climatic change has recently received increased attention, as evidenced by the SMIC report (Wilson, 1971). A leading anthropogenic effect is the enrichment of the atmospheric CO_2 content by the combustion of fossil fuels, which has been rising about 4 percent per year since 1910. There is evidence that the ocean's uptake of much of this CO_2 is diminishing (Keeling et al., 1974), which raises the possibility of even greater future atmospheric concentrations. Man's activities are also contaminating the atmosphere with aerosols and releasing waste heat into the atmosphere, either (or both) of which may have important climatic consequences (Mitchell, 1973b). Such effects may combine to offset a future natural cooling trend or to enhance a natural warming. This situation serves to illustrate the uncertainty introduced into the problem of future climatic changes by the interference of man and is occurring *before* adequate knowledge of the natural variations themselves has been obtained. Again, the clear need is for greatly increased research on both the nature and causes of climatic variation.

REFERENCES

Addicott, W. O., 1970: Latitudinal gradients in Tertiary molluscan faunas of the Pacific coast, *Paleog. Paleoclimatol. Paleocol., 8*:287–312.

Berger, W. H., 1971: Sedimentation of planktonic foraminifera, *Marine Geol., 11*:325–358.

Berggren, W. A., 1972: A Cenozoic time-scale—some implications for regional geology and paleobiogeography, *Lethaia, 5*:195–215.

Bergthorsson, F., 1962: Preliminary notes on past climate of Iceland, Conference on the climate of the 11th and 16th centuries, Aspen, Colo., June 16–24, 1962 (unpublished notes).

Bernabo, J. C., T. Webb, III, and J. McAndrews, 1974: Postglacial isopollen maps of major forest genera and herbs in northeastern North America (in preparation).

Bjerknes, J., 1969a: Atmosphere–ocean interaction during the "Little Ice Age" (seventeenth to nineteenth centuries A.D.), in *WMO Tech. Note No. 66,* 339 pp.

Bjerknes, J., 1969b: Atmospheric teleconnections from the equatorial Pacific, *Mon. Wea. Rev., 97*:163–172.

Bloom, A. L., 1971: Glacial-eustatic and isostatic controls of sea level since the last glaciation, in *The Late Cenozoic Glacial Ages,* K. Turekian, ed., Yale U.P., New Haven, Conn., pp. 355–380.

Brier, G. W., 1968: Long-range prediction of the zonal westerlies and some problems in data analysis, *Rev. Geophys., 6*:525–551.

Broecker, W. S., and A. Kaufman, 1965: Radiocarbon chronology of Lake Lahontan and Lake Bonneville II, Great Basin, *Geol. Soc. Am. Bull., 76*:537–566.

Broecker, W. S., and J. van Donk, 1970: Insolation changes, ice volumes, and the O^{18} record in deep-sea cores, *Rev. Geophys. Space Phys., 8*:169–198.

Bryson, R. A., 1974: The Sahelian effect (to be published in *Ecologist*).

Bryson, R. A., and F. K. Hare, 1973: The climate of North America, in *World Survey of Climatology, 11*, Elsevier, Amsterdam.
Bryson, R. A., W. M. Wendland, J. D. Ives, and J. T. Andrews, 1969: Radiocarbon isochrones on the disintegration of the Laurentide ice sheet, *Arctic Alpine Res., 1*:1–14.
Bryson, R. A., D. A. Baerreis, and W. M. Wendland, 1970: The character of late-glacial and post-glacial climatic changes, in *Pleistocene and Recent Environments of the Central Great Plains*, U. of Kansas Press, Lawrence, Kan., pp. 53–74.
Budyko, M. I., 1963: *Atlas of the Heat Balance of the Globe*, Hydrometeorological Service, Moscow, 69 pp.
Budyko, M. I., 1969: The effect of solar radiation variations on the climate of the earth, *Tellus, 21*:611–619.
Butzer, K. W., G. L. Isaac, J. L. Richardson, and C. Washbourn-Kamau, 1972: Radiocarbon dating of East African lake levels, *Science, 175*:1069–1076.
Crutcher, H. L., and J. M. Meserve, 1970: *Selected level heights, temperatures and dew points for the Northern Hemisphere*, NAVAIR 10–1C–52, Naval Weather Service, Washington, D.C., 370 pp.
Dansgaard, W., 1954: The O^{18} abundance in fresh water, *Geochim. Cosmochim. Acta, 6*:241.
Dansgaard, W. S., S. J. Johnsen, H. B. Clausen, and C. C. Langway, Jr., 1971: Climatic record revealed by the Camp Century ice core, in *The Late Cenozoic Glacial Ages*, K. Turekian, ed., Yale U.P., New Haven, Conn., pp. 37–56.
Denton, G. H., and W. Karlén, 1973: Holocene climatic changes, their pattern and possible cause, *Quaternary Res., 3*:155–205.
Denton, G. H., R. K. Armstrong, and M. Stuiver, 1971: The late Cenozoic glacial history of Antarctica, in *The Late Cenozoic Glacial Ages*, K. Turekian, ed., Yale U.P., New Haven, Conn., pp. 267–306.
Dietz, R. S., and J. C. Holden, 1970: Reconstruction of Pangaea: breakup and dispersion of continents, Permian to present, *J. Geophys. Res., 75*:4939–4956.
Douglas, R. G., and S. M. Savin, 1973: *Initial Reports of the Deep Sea Drilling Project, 17*, U.S. Govt. Printing Office, Washington, D.C., pp. 591–605.
Dreimanis, A., and P. F. Karrow, 1972: Glacial history of the Great Lakes—St. Lawrence Region, the classification of the Wisconsinan stage, and its correlatives, *24th Int. Geol. Congress* (Section 12, Quaternary Geology), pp. 5–15.
Emiliani, C., 1955: Pleistocene temperatures, *J. Geol., 63*:538–578.
Emiliani, C., 1968: Paleotemperature analysis of Caribbean cores P6304–8 and P6304–9 and a generalized temperature curve for the past 425,000 years, *J. Geol., 74*:109–126.
Fletcher, J. O., 1969: Ice extent on the Southern Ocean and its relation to world climate, RM–5793–NSF, The Rand Corporation, Santa Monica, Calif., 119 pp.
Flint, R. F., 1971: *Glacial and Quaternary Geology*, Wiley, New York, 892 pp.
Fritts, H. C., 1965: Tree-ring evidence for climatic changes in western North America, *Mon. Wea. Rev., 93*:421–443.
Fritts, H. C., T. J. Blasing, B. P. Hayden, and J. E. Kutzbach, 1971: Multivariate techniques for specifying tree-growth and climate relationships and for reconstructing anomalies in paleoclimate, *J. Appl. Meteorol., 10*:845–864.
Gilman, D. L., F. J. Fuglister, and J. M. Mitchell, Jr., 1963: On the power spectrum of "red noise," *J. Atmos. Sci., 20*:182–184.

Gommel, W. R., 1963: Mean distribution of 500 mb topography and sea-level pressure in middle and high latitudes of the Northern Hemisphere during the 1950–59 decade, *J. Appl. Meteorol.*, 2:105–113.

Hays, J. D., T. Saito, N. D. Opdyke, and L. H. Burckle, 1969: Pliocene–Pleistocene sediments of the equatorial Pacific, their paleomagnetic biostratigraphic and climatic record, *Bull. Geol. Soc. Am.*, 80:1481–1514.

Holmes, A., 1965: *Principles of Physical Geology*, 2nd ed., Ronald Press, New York, 1288 pp.

Hunkins, K., A. W. H. Bé, N. D. Opdyke, and G. Mathieu, 1971: The late Cenozoic history of the Arctic Ocean, in *The Late Cenozoic Glacial Ages*, K. Turekian, ed., Yale U.P., New Haven, Conn., pp. 215–238.

Imbrie, J., 1972: Correlation of the climatic record of the Camp Century ice core (Greenland) with foraminiferal paleotemperature curves from North Atlantic deep-sea cores, *Geol. Soc. Am.*, Abstracts of 1972 Ann. Mtg., p. 550.

Imbrie, J., and N. G. Kipp, 1971: A new micropaleontological method for quantitative paleoclimatology: application to a late Pleistocene Caribbean core, in *The Late Cenozoic Glacial Ages*, K. Turekian, ed., Yale U.P., New Haven, Conn., pp. 71–182.

Imbrie, J., and J. J. Shackleton, 1974: Climatic periodicities documented by power spectra of the oxygen isotope record in equatorial Pacific deep-sea core V28-238 (in preparation).

Imbrie, J., J. van Donk, and N. G. Kipp, 1973: Paleoclimatic investigation of a late Pleistocene Caribbean deep-sea core: comparison of isotopic and faunal methods, *Quaternary Res.*, 3:10–38.

Iudin, M. I., 1967: On the study of actors determining the nonstationarity of the general circulation, *Proc. Int. Symp. on Dynamics of Large-scale Processes*, Moscow, June 23–30, 1965, Akad. Nauk SSSR, Moscow, 24 pp.

Kaufman, A., 1971: U-Series dating of Dead Sea Basin carbonates, *Geochim. Cosmochim. Acta*, 35:1269–1281.

Keeling, C. D., R. Bacastow, and C. A. Ekdahl, 1974: Diminishing role of the oceans in industrial CO_2 uptake during the next century (in preparation).

Kellogg, T. B., 1974: Late Quaternary climatic changes in the Norwegian and Greenland seas, *Proc. 24th Alaskan Sci. Conf.*, Fairbanks, Alaska, 15–17, 1973 (to be published).

Kennett, J. P., et al., 1973: Deep-sea drilling in the roaring 40's, *Geotimes*, 8:14–17.

Kipp, N. G., 1974: A new transfer function for estimating past sea-surface conditions from the sea-bed distribution of planktonic foraminiferal assemblages, *Geol. Soc. Am. Spec. Paper* (in press).

Kraus, E. B., 1955a: Secular changes of east-coast rainfall regimes, *Q. J. R. Meteorol. Soc.*, 81:430–439.

Kraus, E. B., 1955b: Secular changes of tropical rainfall regimes, *Q. J. R. Meteorol. Soc.*, 81:198–210.

Krueger, A. F., J. S. Winston, and D. A. Haines, 1965: Computation of atmospheric energy and its transformation for the Northern Hemisphere for a recent five-year period, *Mon. Wea. Rev.*, 93:227–238.

Kukla, G. J., 1970: Correlations between loesses and deep-sea sediments, *Geol. Fören, Stockholm Förh.*, 92:148–180.

Kukla, G. J., and H. J. Kukla, 1974: Increased surface albedo in the Northern Hemisphere, *Science*, 183:709–714.

Kukla, G. J., and R. K. Matthews, 1972: When will the present interglacial end?, *Science, 178*:190–191.
Kung, E. C., and S. Soong, 1969: Seasonal variation of kinetic energy in the atmosphere, *Q. J. R. Meteorol. Soc., 95*:501–512.
Kutzbach, J. E., 1970: Large-scale features of monthly mean Northern Hemisphere anomaly maps of sea-level pressure, *Mon. Wea. Rev., 98*:708–716.
Kutzbach, J. E., and R. A. Bryson, 1974: Variance spectrum of Holocene climatic fluctuations in the North Atlantic sector, Dept. of Meteorol., U. of Wisconsin, Madison (unpublished).
LaMarche, V. C., Jr., 1974: Paleoclimatic inferences from long tree-ring records, *Science, 183*:1043–1048.
Lamb, H. H., 1966: Climate in the 1960s, *Geogr. J., 132*:183–212.
Lamb, H. H., 1969: Climatic fluctuations, in *World Survey of Climatology, 2, General Climatology,* H. Flohn, ed., Elsevier, New York, pp. 173–249.
Lamb, H. H., 1972: *Climate: Present, Past and Future,* Vol. 1, Methuen, London, 613 pp.
Lamb, H. H., and A. I. Johnson, 1959: Climatic variation and observed changes in the general circulation (Parts I, II), *Geogr. Ann., 41*:94–134.
Lamb, H. H., and A. I. Johnson, 1961: Climatic variation and observed changes in the general circulation (Part III), *Geogr. Ann., 43*:363–400.
Lamb, H. H., and A. I. Johnson, 1966: Secular variation of the atmospheric circulation since 1750, *Geophys. Mem., 110,* Meteorol. Office, London, 125 pp.
Lamb, H. H., and R. A. S. Ratcliffe, 1972: On the magnitude of climatic anomalies in the oceans and some related observations of atmospheric circulation behavior, in *Climate: Present, Past and Future, 1,* H. H. Lamb, ed., Methuen, London, 613 pp.
Lamb, H. H., R. P. W. Lewis, and A. Woodroffe, 1966: Atmospheric circulation and the main climatic variables between 8000 and 0 B.C.: meteorological evidence, in *World Climate Between 8000 and 0 B.C.,* Royal Meteor. Soc., London, pp. 174–217.
LeRoy Ladurie, E., 1967: *Histoire du Climat depuis l'An Mille,* Flammarion, Paris.
Manley, G., 1959: Temperature trends in England, 1680–1959, *Arch. Meteorol. Geophys. Bioklimatol., Ser. B, 9*:413–433.
Matthews, R. K., 1973: Relative elevation of late Pleistocene high sea level stands: Barbados uplift rates and their implications, *Quaternary Res., 3*:147–153.
Mayewski, P. A., 1973: Glacial geology and late Cenozoic history of the Transantarctic Mountains, Antarctica, PhD Dissertation, Ohio State U., Columbus, Ohio, 216 pp. (unpublished).
McIntyre, A., 1974: Spatial pattern of climate during a glacial age 18,000 years ago: the world-ocean paleoisotherm map, *Program of the 55th annual meeting of the Am. Geophys. Union* (abstract).
McIntyre, A., W. F. Ruddiman, and R. Jantzen, 1972: Southward penetrations of the North Atlantic polar front; faunal and floral evidence of large-scale surface-water mass movements over the last 225,000 years, *Deap-Sea Res., 19*:61–77.
McIntyre, A., et al., 1974: The glacial North Atlantic 18,000 years ago: a CLIMAP reconstruction, *Geol. Soc. Am. Spec. Paper* (in press).
Mesolella, K. J., R. K. Matthews, W. S. Broecker, and D. L. Thurber, 1969: The astronomic theory of climatic change: Barbados data, *J. Geol., 77*:250–274.

Milliman, J. D., and K. O. Emery, 1968: Sea levels during the past 35,000 years, *Science, 162*:1121–1123.

Mitchell, J. M., Jr., 1963: On the world-wide pattern of secular temperature change, in *Changes of Climate,* Arid Zone Research XX, UNESCO, Paris, pp. 161–181.

Mitchell, J. M., Jr., 1973a: The natural breakdown of the present interglacial and its possible intervention by human activities, *Quaternary Res., 2*:436–445.

Mitchell, J. M., Jr., 1973b: A reassessment of atmospheric pollution as a cause of long-term changes of global temperature, in *Global Effects of Environmental Pollution,* 2nd ed., S. F. Singer, ed., Reidel, Dordrecht, Holland.

Moore, T. C., Jr., 1972: Successes, failures, proposals, *Geotimes, 17*(7):27–31.

Namias, J., 1969: Seasonal interactions between the North Pacific Ocean and the atmosphere during the 1960's, *Mon. Wea. Rev., 97*:173–192.

Namias, J., 1970: Climatic anomaly over the United States during the 1960's, *Science, 170*:741–743.

Namias, J., 1972a: Influence of Northern Hemisphere general circulation on drought in northeast Brazil, *Tellus, 24*:336–343.

Namias, J., 1972b: Large-scale and long-term fluctuations in some atmospheric and oceanic variables, in *The Changing Chemistry of the Oceans, Nobel Symposium 20,* O. Dryssen and D. Jagner, eds., Wiley, New York, pp. 27–48.

O'Connor, J. F., 1969: Hemispheric teleconnections of mean circulation anomalies at 700 millibars, *ESSA Tech. Rep. WB 10,* U.S. Govt. Printing Office, Washington, D.C., 103 pp.

Oort, A. H., and E. M. Rasmusson, 1971: Atmospheric circulation statistics, *NOAA Prof. Paper 5,* U.S. Govt. Printing Office, Washington, D.C., 323 pp.

Petterssen, S., 1969: *Introduction to Meteorology,* McGraw-Hill, New York, 338 pp.

Porter, S. C., and G. H. Denton, 1967: Chronology of Neoglaciation in the North American Cordillera, *Am. J. Sci., 265*:117–119.

Prell, W. L., 1974: Late Pleistocene faunal and temperature patterns of the Columbia Basin, Caribbean Sea, in *Geol. Soc. Am. Spec. Paper* (in press).

Reitan, C. H., 1971: An assessment of the role of volcanic dust in determining modern changes in the temperature of the Northern Hemisphere, PhD Thesis, Dept. of Meteorol., U. of Wisconsin, Madison, Wisc., 147 pp.

Sancetta, C., J. Imbrie, and N. G. Kipp, 1973: Climatic record of the past 130,000 years in North Atlantic deep-sea core V23-82: correlation with the terrestrial record, *Quaternary Res., 3*:110–116.

Sawyer, J. S., 1964: Notes on the possible physical causes of long-term weather anomalies, in WMO-IUGG Symposium on Research and Development Aspects of Long-Range Forecasting, *WMO Tech. Note No. 66,* 339 pp.

Servant, M., *et al.,* 1969: Chronologie du Quaternaire récent des basses régions du Tchad, *C.R. Acad. Sci. Paris, 269*:1603–1606.

Shackleton, N. J., and J. P. Kennett, 1974a: Oxygen and carbon isotope record at DSDP Site 284 and their implications for glacial development, *Initial Reports of the Deep Sea Drilling Project, 29* (in press).

Shackleton, N. J., and J. P. Kennett, 1974b: Paleotemperature history of the Cenozoic and the initiation of Antarctic glaciation; oxygen and carbon isotope analyses in DSDP Sites to 77, 279, and 281, *Initial Reports of the Deep Sea Drilling Project, 29* (in press).

Shackleton, N. J., and N. D. Opdyke, 1973: Oxygen isotope and paleomagnetic stratigraphy of equatorial Pacific core V28-238: oxygen isotope temperatures and ice volumes on a 10^5 and 10^6 year scale, *Quaternary Res.*, *3*:39-55.

Starr, V. P., and A. H. Oort, 1973: Five-year climatic trend for the Northern Hemisphere, *Nature*, *242*:310-313.

Steiner, J., and E. Grillmair, 1973: Possible galactic causes for periodic and episodic glaciation, *Geol. Soc. Am. Bull.*, *84*:1003-1018.

Suess, H. E., 1970: The three causes of secular C^{14} fluctuations, their amplitudes and time constants, in *Radiocarbon Variations and Absolute Chronology, Nobel Symposium 12,* I. U. Olsson, ed., Wiley, New York, pp. 595-604.

Sverdrup, H. U., M. W. Johnson, and R. H. Fleming, 1942: *The Oceans*, Prentice-Hall, Englewood Cliffs, N.J., 1087 pp.

Taljaard, J. J., H. Van Loon, H. L. Crutcher, and R. L. Jenne, 1969: *Climate of the Upper Air: Southern Hemisphere, 1.* Temperatures, dew points and heights at selected pressure levels, NAVAIR 50-1C-55, Naval Weather Service, Washington, D.C.

Van der Hammen, T., T. A. Wijmstra, and W. H. Zagwijn, 1971: The floral record of the late Cenozoic of Europe, in *The Late Cenozoic Glacial Ages,* K. Turekian, ed., Yale U.P., New Haven, Conn., pp. 391-424.

Veeh, H. H., and J. Chappell, 1970: Astronomical theory of climatic change: support from New Guinea, *Science*, *167*:862-865.

Vonder Haar, T. H., and A. H. Oort, 1973: New estimates of annual poleward energy transport by Northern Hemisphere oceans, *J. Phys. Oceanog.*, *3*:169-172.

Vonder Haar, T. H., and V. E. Suomi, 1971: Measurements of the earth's radiation budget from satellites during a five-year period, Part I: Extended time and space means, *J. Atmos. Sci.*, *28*:305-314.

Wagner, A. J., 1971: Long-period variations in seasonal sea-level pressure over the Northern Hemisphere, *Mon. Wea. Rev.*, *99*:49-69.

Wahl, E. W., 1972: Climatological studies of the large-scale circulation in the Northern Hemisphere, *Mon. Wea. Rev.*, *100*:553-564.

Walcott, R. I., 1972: Past sea levels, eustasy and deformation of the earth, *Quaternary Res.*, *2*:1-14.

Washburn, A. L., 1973: *Periglacial Processes and Environments,* Arnold Press, London, 320 pp.

Webb, T., and R. A. Bryson, 1972: Late- and post-glacial climatic change in the northern midwest, USA: quantitative estimates derived from fossil pollen spectra by multivariate statistical analysis, *Quaternary Res.*, *2*:70-111.

Weyl, P. K., 1968: The role of he oceans in climatic change: a theory of the ice ages, *Meteorol. Monogr.*, *8*:37-62.

Willett, H. C., 1967: Maps of standard deviation of monthly mean sea level pressure for January, April, July and October, 1899-1960, Massachusetts Inst. of Tech., Cambridge, Mass. (unpublished).

Wilson, C. L. (Chairman), 1971: Study of Man's Impact on Climate (SMIC) Report, *Inadvertent Climate Modification,* W. H. Matthews, W. W. Kellogg, and G. D. Robinson, eds., MIT Press, Cambridge, Mass., 308 pp.

Winstanley, D., 1973: Rainfall patterns and general atmospheric circulation, *Nature*, *245*:190-194.

Winston, J. S., 1969: Temporal and meridional variations in zonal mean radiative heating measured by satellites and related variations in atmospheric energetics, PhD dissertation, Dept. of Meteorol. and Oceanog., NYU, New York, 152 pp.

APPENDIX B
SURVEY OF THE CLIMATE SIMULATION CAPABILITY OF GLOBAL CIRCULATION MODELS

INTRODUCTION

Much of the present effort within GARP, as well as other research programs in the atmospheric and oceanic sciences, is aimed toward the development of a quantitative understanding of the behavior of the atmosphere, with the immediate objective of improving the accuracy of weather forecasts. Other research efforts and plans, and the research program proposed in this report in particular, are directed to the longer-range objective of understanding the physical basis of climate and climatic change. Essential to both of these objectives are the dynamical models of the global atmospheric and oceanic circulation.

These general circulation models (or GCM's) have been developed over a number of years, in parallel with the growth of computing capability and the increase of atmospheric data coverage. The several atmospheric and oceanic GCM's have now reached the point where reasonably accurate simulations of the global distribution of many important climatic elements are possible and where their coupling into a single dynamical system is now feasible. This therefore seems to be a useful time to survey briefly these models' climate simulation capabilities.

Here we have not attempted to present a detailed discussion of the various GCM's, as such descriptions are readily available both in the literature and in documents describing special models. Model reviews have recently been prepared by Robinson (1971), Willson (1973), Smagorinsky (1974), and Schneider and Dickinson (1974), and general discussions of the use of such models for weather prediction and for studies of the general circulation are available [see, for example, the review by Smagorinsky (1970) and also Haltiner (1971) and Lorenz (1967)]. A survey of the physical and mathematical structure of both regional and global atmospheric models is also in preparation for GARP (1974). What has not been assembled heretofore is the comparative climatic performance of the various models, and this Appendix is an initial effort to fill this need for both the atmospheric and oceanic global GCM's.

In general, any formulation that relates variables of the climatic system to the external or boundary conditions may be considered a climatic *model*. We can thus identify basically empirical and statistical climatic models, as well as those that rest on the system's dynamical equations.

Within the dynamical climate models, a wide variety of the type and degree of parameterization may be seen. At one extreme are the vertically and zonally averaged atmospheric models that address the mean heat balance at the earth's surface, such as those of Budyko (1969) and Sellers (1973). In such models, the transport of heat is parameterized in terms of mean zonal variables, which are in turn related to the surface temperature. At the other extreme are the high-resolution global general circulation models or GCM's. In these models, the details of the transient cyclone-scale motions are resolved, along with the global distribution of the elements of the heat and hydrologic balances. Even these models, however, parameterize certain physical processes, in that they employ empirical or statistical representations of some of the subgrid scale processes in the surface boundary layer and in the free atmosphere and open ocean, such as the effects of diffusion and convection.

Dynamical climate models also display a wide variety of parameterization with respect to time. This ranges from equilibrium or steady-state models, such as that of Saltzman and Vernekar (1971), to the GCM's that explicitly calculate the time dependence of the circulation in steps of a few minutes. With respect to their treatment of both space and time, therefore, a wide range of models exists, and each is suited to the investigation of particular aspects of the climatic problem. The GCM's (of both the atmosphere and ocean) provide the most detailed representation of the physical processes involved but require large amounts of computation. These models have therefore been used up to the present time to study only the climatic variations on time scales of the order of years (for the atmosphere) to centuries (for the oceans). The more highly parameterized models, on the other hand, provide less detail but are capable of treating the longer-period climatic variations with much less computation. Once they are adequately calibrated with respect to observations, an important use of the GCM's will be to generate detailed climatic statistics, from which parameterizations appropriate to the various statistical–dynamical models may be prepared.

In the remainder of this Appendix we give our attention to the principal atmospheric and oceanic general circulation models, for the purpose of indicating their present capability to simulate climate. Before presenting these results, however, it is useful to review briefly the historical development of numerical modeling in general.

DEVELOPMENT AND USES OF NUMERICAL MODELING

The basis for the mathematical modeling of the behavior of the atmosphere was first unambiguously stated by V. Bjerknes in 1904. It is only in the last 20 years or so, however, that the means for carrying out such modeling on a practical basis have become available. These include adequate observations for model calibration and verification, a knowledge of the important physical processes and their parameterization, and the computers and numerical methods necessary to perform the calculations.

The observational base for numerical modeling of the atmosphere has grown steadily since the 1940's and early 1950's, when the global radiosonde network began to take shape. The IGY provided further expansion, but the observational coverage still needs augmentation, especially over the oceanic regions. The real breakthrough toward the global measurements necessary for numerical modeling has come from the remote-sensing capabilities of meteorological satellites; with the aid of suitable surface (ground-truth) observations, these are capable of providing the first truly worldwide observations of the air and ocean surface temperature, moisture and cloudiness, and elements of the heat and hydrologic balance. By using the numerical models diagnostically, there is then the prospect of deducing the accompanying global distributions of other variables, such as the wind velocity. Such a scheme is the observational basis of the proposed First GARP Global Experiment (FGGE) in 1978.

The physical and theoretical basis for numerical modeling has grown significantly with the development of the theory of baroclinic instability, the parameterization of moist convection, and advances in our knowledge of the behavior of the stratosphere and the planetary boundary layer. Our growing understanding of these processes has increased the prospects for improved weather forecasts. These hopes are bounded, however, by the realization that the atmosphere possesses limited predictability, i.e., that there is a time range beyond which the local variations of weather appear as random fluctuations as far as their explicit prediction by numerical models is concerned. Present indications are that this limit lies at about two weeks' time.

The key physical processes that control the longer-period variations of the atmosphere—those that are properly associated with climate—are largely unknown, although we are beginning to recognize the importance of a number of feedback relationships, such as the air–sea coupling and cloudiness–temperature feedback. Numerical models that incorporate such effects are our best tool to develop a quantitative understanding of their role in climate and climatic variation.

The computational base for numerical modeling has grown during the last 20 years in parallel with the development of successive generations of high-speed computers, as shown in Figure B.1. This overview makes clear the interrelated development of numerical models, theory, and computer speed. Numerical weather prediction may be considered to have begun with the first successful numerical integration of the

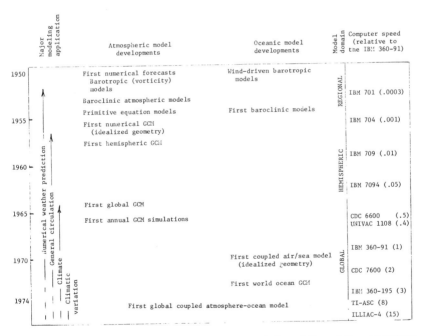

FIGURE B.1 Highlights in the development of numerical modeling of the atmosphere and ocean.

vorticity equation (Charney et al., 1950), with the demonstration of the ability of baroclinic models to forecast cyclonic development (Charney and Phillips, 1953), or with the commencement of operational numerical weather prediction in 1955. Numerical general circulation studies may be considered to have begun with the simulation of the atmospheric energy cycle in an idealized model with sources and sinks of energy and momentum (Phillips, 1956), with the first successful hemispheric circulation experiments (Smagorinsky, 1963), or with the first extended global integration (Mintz, 1965). Numerical climate models for the atmosphere may be considered to have begun with the global simulation of the seasonal and interannual variation of the primary climatic elements (Mintz et al., 1972), although the modeling of climate by other methods has a much longer history. The numerical modeling of climatic variation, on the other hand, which addresses the coupled ocean–atmosphere climatic system, has only just begun (Bryan et al., 1974; Manabe et al., 1974a, 1974b).

The development during the past decade of numerical methods whose stability and accuracy can be suitably controlled has made it possible to carry out such calculations for extended periods of time. Even with today's fastest computers, however, the solution of the more detailed global numerical models proceeds only at a rate between one and two orders of magnitude faster than nature itself, and our ability to perform the large number of numerical integrations required for the systematic exploration of climate and climatic change requires the continued development and dedication of new computer resources.

A similar pattern of development has occurred in the numerical modeling of the oceans, except that the rate of progress has been slower due principally to a lack of suitable oceanic observations. The data base for the oceans is fragmentary in comparison with that for the atmosphere, and there is no oceanic counterpart of the radiosonde or weather station network. The bathythermograph has been widely used to measure the thermal structure of the ocean's surface layer for the past few decades, but even this has not been done on a synoptic basis. The bulk of the data for oceanic temperature, salinity, and currents has been obtained in the course of occasional oceanographic expeditions or special observational programs. Even so, the number of direct velocity measurements is quite small, and our knowledge of the oceanic circulation is largely based on geostrophic estimates from conventional hydrographic observations.

Our knowledge of the dynamics of the ocean circulation is also less complete than is that for the atmosphere. While the character of the vorticity balance of the ocean was first established by Sverdrup (1947) and Stommel (1948), the role of the thermohaline circulation was demonstrated with a numerical model only a few years ago (Bryan and Cox, 1968), and the effects of bottom topography have been established even more recently (see, for example, Holland and Hirschman, 1972). Numerical models are proving of great value in the study of time-dependent behavior of the oceanic general circulation and in the analysis of oceanic mesoscale motions such as those now being revealed by the MODE observations. The structure of these eddies and the role that they play in the oceanic heat balance is one of the principal unsolved problems in physical oceanography. Other important questions concern the nature of vertical mixing in the ocean, especially in the surface layer, and the mechanics of the formation of deep and bottom water. Each of these can perhaps be most fruitfully studied with appropriate regional numerical models, in order to lay the foundation for their parameterization in three-dimensional models of the world ocean. But perhaps the most important problem of all from the viewpoint of climate is the *interaction* between the ocean and the atmosphere; the numerical modeling of this coupled system offers our best hope of achieving a quantitative understanding of the dynamics of climatic variation.

Numerical models thus lie at the heart of the modern study of climate and climatic change: they complement (and may even be regarded as a part of) the observing system, they serve as tools for climatic analysis and diagnosis, and they offer the most rational way of assessing the course of future climatic events. Whether or not climate forecasting in the time-dependent sense ever becomes feasible, the use of numerical models to simulate the average or equilibrium climates of the past and the likely climatic consequences of various natural or anthropogenic effects in the future will justify their development.

ATMOSPHERIC GENERAL CIRCULATION MODELS

Formulation

All general circulation models are based on the fundamental dynamical equations that govern the large-scale behavior of the atmosphere. This system consists of the equation of motion (expressing the conservation of momentum), the thermodynamic energy equation (expressing the conservation of heat energy), the equations of mass and water vapor continuity, and the equation of state. When geometric height (z) is the vertical coordinate, these equations can be written in vector form as follows:

$$\frac{\partial \vec{V}}{\partial t} + \vec{V} \cdot \nabla \vec{V} + w \frac{\partial \vec{V}}{\partial z} + 2\vec{\Omega} \times \vec{V} + \frac{1}{\rho} \nabla p = \vec{F}, \quad (1)$$

$$\frac{\partial p}{\partial z} + \rho g = 0, \quad (2)$$

$$\frac{\partial \theta}{\partial t} + \vec{V} \cdot \nabla \theta + w \frac{\partial \theta}{\partial z} = Q, \quad (3)$$

$$\frac{\partial \rho}{\partial t} + \nabla \cdot \rho \vec{V} + \frac{\partial}{\partial z}(\rho w) = 0, \quad (4)$$

$$\frac{\partial q}{\partial t} + \vec{V} \cdot \nabla q + w \frac{\partial q}{\partial z} = S, \quad (5)$$

$$p = \rho R T. \quad (6)$$

Here $\vec{V}$ is the horizontal velocity, w is the vertical velocity, $\vec{\Omega}$ is the rotation vector of the earth, ρ is the density, p is the pressure, g is the gravitational acceleration, θ is potential temperature [which is related to the ordinary temperature T by the relation $\theta = T(p_0/p)^\kappa$, where $p_0 = 1000$ mbar and $\kappa = 0.286$ is the ratio of the specific heats], q is the water vapor mixing ratio, R is the gas constant for (moist) air, and ∇ is the horizontal gradient operator.

The terms $\vec{F}$, Q, and S on the right-hand sides of Eqs. (1), (3), and (5) represent the sources and sinks of momentum, heat, and water vapor due to a variety of physical processes in the atmosphere and must be either prescribed or parameterized in terms of the primary dependent variables in order to close the system (1)–(6). The net frictional force $\vec{F}$ consists of the frictional drag at the earth's surface and the internal friction in the free atmosphere, as well as the changes of large-scale momentum due to smaller-scale processes. The net (diabatic) heating rate Q consists of the latent heat released during condensation, the heating due to the exchange of both long-wave and shortwave radiation, and the sensible heating of the atmosphere by turbulent heat fluxes from the underlying surface. The net moisture addition rate S consists of the difference between the evaporation rate (from both the surface and from cloud and precipitation) and the condensation rate.

An important contribution to each of these source terms is the vertical flux of momentum, heat, and moisture, which accompanies cumulus-scale convection in the atmosphere. We may note that such convective-scale processes are not governed by the system (1)–(6) and must be represented in terms of the larger-scale variables. This parameterization is particularly critical for the net heating, because most of the latent heating in the atmosphere is accomplished by convective motions, which are also responsible for much of the cloudiness (see Figure 3.2).

The various atmospheric GCM's are each formulated in slightly different ways and employ different treatments of the source terms. There is at present insufficient evidence to decide which particular formulation is the most satisfactory, and there is even more uncertainty regarding the most correct parameterization of the unresolved physical processes contained within $\vec{F}$, Q, and S. A summary of some of the features of the better-known atmospheric general circulation models is given in Table B.1. Each of the models shown here uses generally similar procedures to determine the ground-surface temperature (from an assumed heat balance over land and ice), the surface hydrology (with runoff permitted after saturation of the surface soil), and the occurrence of convection (from vertical stability criteria depending on the moist static energy). Each of the models also incorporates the observed large-scale distributions of terrain height, surface albedo, and sea-surface temperature.

Solution Methods

All the atmospheric GCM's considered here employ finite-difference methods of second-order accuracy, with the dependent variables generally determined on a spatially staggered grid with a resolution of several hundred kilometers (see Table B.1). Time differencing is also generally of second-order accuracy, with time steps between 5 and 10 min used to maintain (linear) computational stability. Long-term (nonlinear) computational stability is inherent in some of the models' space differencing schemes, while others employ eddy diffusion processes to achieve this end. Various degrees of smoothing are also employed in the models' solution, in addition to that inherent in the finite-difference approximations themselves. Depending on the computer used, the number of model levels, and the frequency with which the radiative heating calculations are performed, global atmospheric GCM's generally run between 10 and 100 times faster than real time.

Selected Climatic Simulations

In order to display the level of accuracy characteristic of present-day atmospheric GCM's in the simulation of climate, we have here assembled the results of model integrations drawn from recently published (and in some cases as yet unpublished) sources. To facilitate comparison, these are presented in a common format, along with the corresponding observed distributions.

TABLE B.1 Principal Characteristics of Current Global Atmospheric General Circulation Models

	GFDL [a]	GFDL [b]	NCAR [c]	UCLA [d]	Rand [e]	GISS [f]
Number of levels and vertical coordinate	9(σ)	11(σ)	6, 12(z)	3, 12(σ)	2(σ)	9(σ)
Model top (mbar)	16	10	70, 5	100, 1	200	10
Horizontal resolution	550 km	250 km	$\Delta\phi=5°$, 2½° $\Delta\lambda=5°$, 2½°	$\Delta\phi=4°$ $\Delta\lambda=5°$	$\Delta\phi=4°$ $\Delta\lambda=5°$	$\Delta\phi=4°$ $\Delta\lambda=5°$
Time step (min)	10	5	6	6	6	5
Insolation	Annual mean	Seasonal	Seasonal	Seasonal	Seasonal	Seasonal
Cloudiness	Specified	Specified	Variable	Variable	Variable	Variable
Lateral eddy viscosity	Nonlinear	Nonlinear	Nonlinear	None	None	None
Sea-surface temperature	Predicted	Assigned	Assigned	Assigned	Assigned	Assigned
Snow cover	Predicted	Predicted	Predicted	Predicted	Assigned	Assigned
Evaporation of falling precipitation	No	No	No	Yes	No	Yes
Sea ice	Predicted	Predicted	Assigned	Assigned	Assigned	Assigned

[a] A coupled ocean–atmosphere model (Manabe et al., 1974a; Bryan et al., 1974).
[b] Manabe et al. (1974b).
[c] Kasahara and Washington (1971).
[d] Arakawa and Mintz (1974).
[e] Gates (1972).
[f] Somerville et al. (1974).

Sea-Level Pressure

Although the various GCM's differ greatly in their resolution of the vertical structure of the atmosphere, each simulates the distribution of a number of climatic variables at the earth's surface. Of these, perhaps the distribution of sea-level pressure is the most familiar; it is shown here as simulated by four different models for the month of January.

In Figure B.2 the average sea-level pressure simulated by the 11-level GFDL atmospheric model is shown for the months of December, January, and February (Manabe et al., 1974b). Figures B.3, B.4, and B.5 show the corresponding average January sea-level pressure simulated by the six-level NCAR model (Kasahara and Washington, 1971), by the two-level Rand model (Gates, 1972), and by the nine-level GISS model (Somerville et al., 1974). In each case the observed average January sea-level pressure distribution is also shown. While the models' results differ in a number of details, these results generally show a useful level of accuracy. As might be anticipated, the largest errors (and the greatest differences among the models) occur in the middle and higher latitudes of the northern hemisphere where cyclonic activity is the most frequent. It should be recalled, however, that sea-level pressure alone is by no means a complete indicator of climate.

Tropospheric Temperature and Pressure

In Figure B.6 the average January 800-mbar temperature simulated by the two-level Rand model (Gates, 1972) is shown, along with the observed distribution. Although systematic errors may be noted over the continents, the simulated large-scale temperature distribution clearly reflects the positions of the major thermal perturbations in the lower troposphere. The average January 500-mbar height simulated by the nine-level GISS model (Somerville et al., 1974) is shown in Figure B.7, along with the observed distribution. These results also clearly show that the mean position and intensity of the long waves in the westerlies are portrayed reasonably well in the simulation.

Cloudiness and Precipitation

Among the more difficult climatic elements to simulate accurately in a GCM are the cloudiness and precipitation. This is doubtless due to the fact that a substantial portion of the total cloudiness and precipitation observed occurs in connection with convective-scale motions, especially

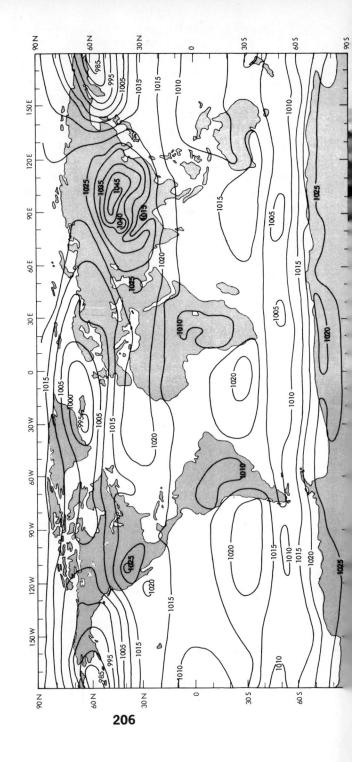

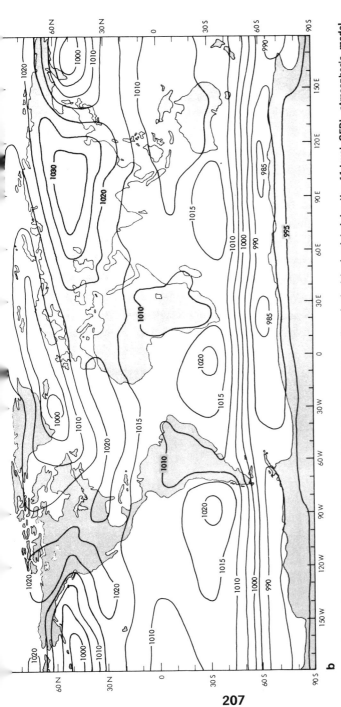

FIGURE B.2 The December–January–February average sea-level pressure (in mbar): (a) simulated by the 11-level GFDL atmospheric model (redrawn from Manabe et al., 1974b); (b) observed January from Schutz and Gates (1971), based on data of Crutcher and Meserve (1970) and Taljaard et al. (1969).

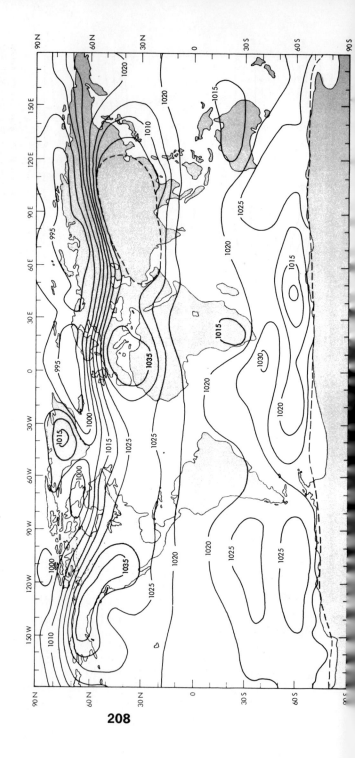

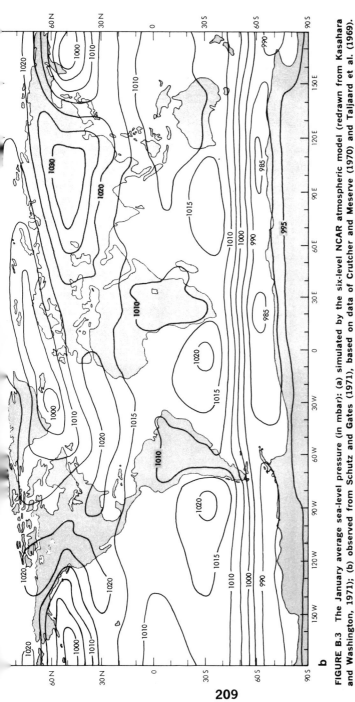

FIGURE B.3 The January average sea-level pressure (in mbar): (a) simulated by the six-level NCAR atmospheric model (redrawn from Kasahara and Washington, 1971); (b) observed from Schutz and Gates (1971), based on data of Crutcher and Meserve (1970) and Taljaard et al. (1969).

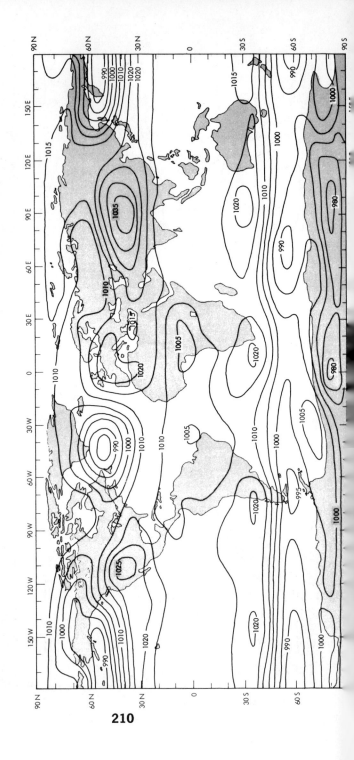

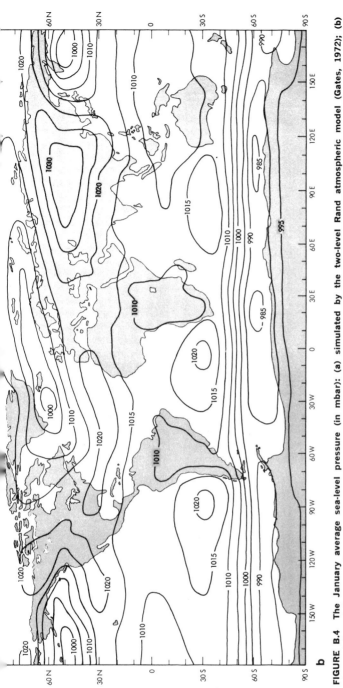

FIGURE B.4 The January average sea-level pressure (in mbar): (a) simulated by the two-level Rand atmospheric model (Gates, 1972); (b) observed from Schutz and Gates (1971), based on data of Crutcher and Meserve (1970) and Taljaard et al. (1969).

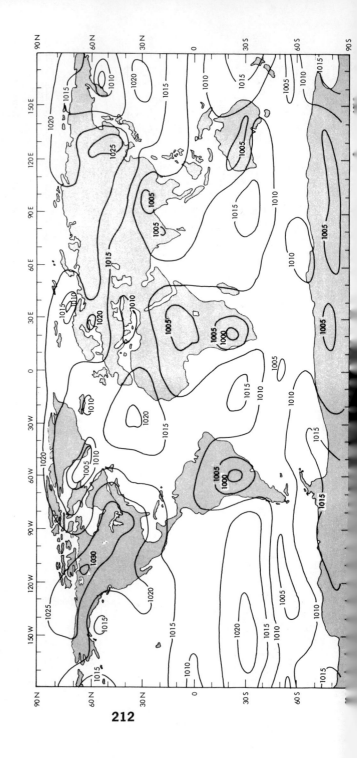

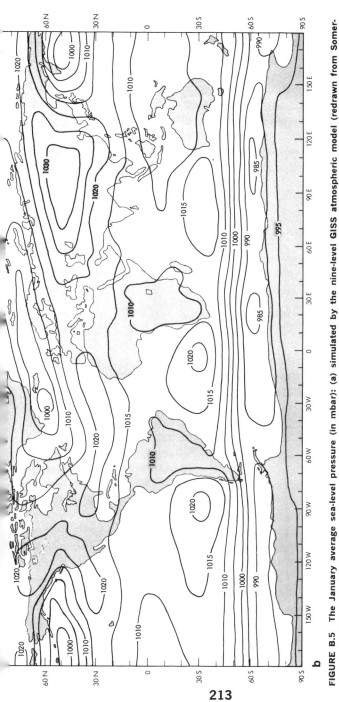

FIGURE B.5 The January average sea-level pressure (in mbar): (a) simulated by the nine-level GISS atmospheric model (redrawn from Somerville et al., (1974); (b) observed from Schutz and Gates 1971), based on data of Crutcher and Meserve (1970) and Taljaard et al. (1969).

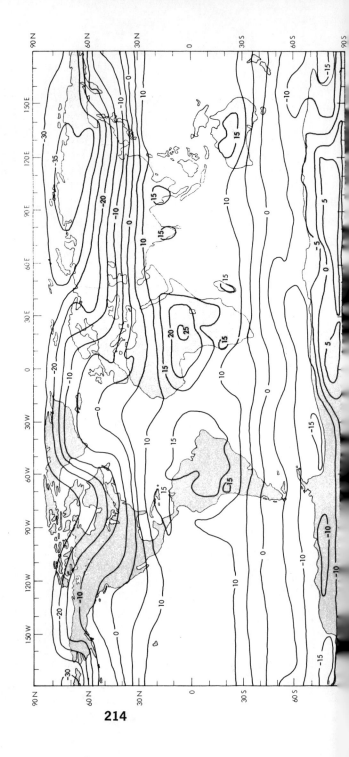

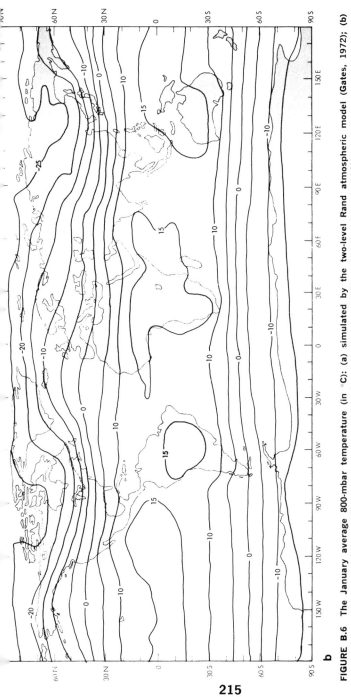

FIGURE B.6 The January average 800-mbar temperature (in °C): (a) simulated by the two-level Rand atmospheric model (Gates, 1972); (b) observed from Schutz and Gates (1971), based on data of Crutcher and Meserve (1970), and Taljaard et al. (1969).

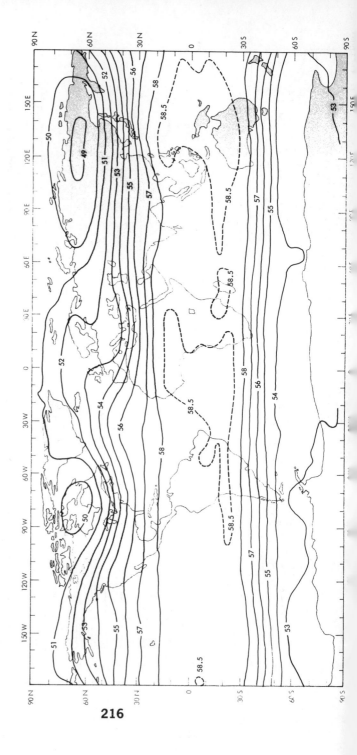

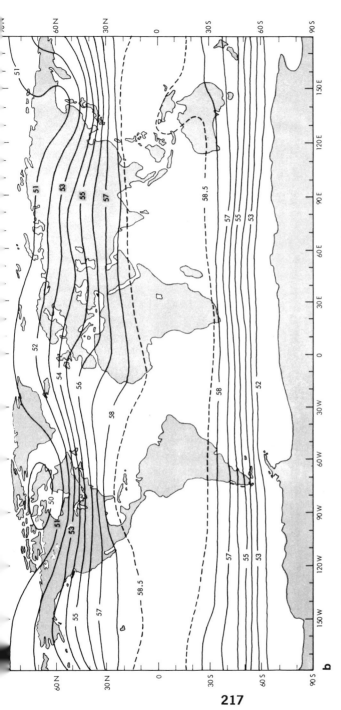

FIGURE B.7 The January average 500-mbar height (in 10^2 m): (a) simulated by the nine-level GISS atmospheric model (redrawn from Somerville et al., 1974); (b) observed from Heastie and Stephenson (1960).

in the lower latitudes. As noted earlier, these processes must be parameterized in the GCM's, and their accurate calibration is relatively difficult.

In Figure B.8, the average January total cloudiness simulated by the six-level NCAR model (Kasahara and Washington, 1971) is shown, along with a composite observed distribution for January and for December, January, and February. With the exception of the equatorial region and the low latitudes of the northern hemisphere, the large-scale areas of maximum and minimum cloudiness are reasonably well simulated.

In Figure B.9, the annual average precipitation simulated by the 11-level GFDL model (Manabe et al., 1974b) is shown, along with the corresponding observed distribution. In addition to the large-scale precipitation pattern in middle latitudes, this simulation also portrays a number of the smaller-scale features, including the zone of heavy precipitation near the equator. Although this comparison is for a somewhat longer time period than the others shown here, the difficulty of correctly parameterizing the precipitation process makes the skill of this simulation impressive.

OCEANIC AND COUPLED ATMOSPHERE–OCEAN GENERAL CIRCULATION MODELS

Estimates based on observed data show that the heat transported by ocean currents plays a major role in the global heat balance (Vonder Haar and Oort, 1973). A model that is to be useful for the study of climatic variation must therefore include the ocean as well as the atmosphere. As suggested by the simulations just reviewed, the specification of a fixed ocean surface temperature in atmospheric GCM's is a strong boundary condition and may mask weaknesses in the models' simulation of the heat balance. The problem of climatic variation therefore furnishes a major motivation for the accelerated development of numerical models of the oceanic general circulation.

Relative to numerical models of the atmosphere, numerical modeling of the ocean is still in a primitive state. As previously noted, this is primarily due to the lack of sufficient data to perform a careful verification of the models and to parameterize properly the effects of the smaller-scale motions. The only large body of data presently available for verifying ocean circulation models is the collection of measurements of density structure. While these data were sufficient to calibrate the earlier analytic theories of the ocean thermocline, global numerical models require a much more extensive data base for adequate verification.

APPENDIX B   219

It is now recognized that many of the earlier studies, such as those by Bryan and Cox (1967) and Haney (1974) for idealized basins, as well as the higher-resolution simulations of Cox (1970) for the Indian Ocean and of Friedrich (1970) for the North Atlantic, represent transient rather than equilibrium solutions for the boundary conditions imposed. The extended integration of even more detailed numerical models will be necessary in the future, in order to design and calibrate adequately other simpler models. Such models will require less calculation and thereby allow more freedom to carry out the large number of numerical experiments required. General reviews of numerical modeling of the ocean circulation are given in the proceedings of a recent symposium (Ocean Affairs Board, 1974) and by Gilbert (1974).

Formulation

The principal dynamical components of an oceanic general circulation model are similar to those of its atmospheric counterpart, namely, the equations of motion, conservation equations for potential temperature and salinity, the continuity equation, and an equation of state. In addition, an oceanic model should contain equations for the growth and movement of pack ice.

In some problems of oceanic circulation, it is not necessary to treat the temperature and salinity separately, and these variables can be combined into a single density variable. In climatic studies, however, we are interested in the *heat* transported by ocean currents explicitly; and in many regions of the world ocean, particularly the polar seas, the density and temperature are not proportional. In these regions at least, it is therefore necessary to predict salinity as a separate independent variable. A changing salinity structure in the ocean may provide the basis of climatic change mechanisms that have not yet received sufficient attention.

In an ocean model, the equation of motion (1) may be simplified by treating the density ρ as a constant ρ_0 (Boussinesq approximation), while the hydrostatic equation (2) remains unchanged. The thermodynamic energy equation (3) and the water vapor continuity equation (5) are represented in the ocean by conservation equations for potential temperature θ and salinity s of the form

$$\frac{\partial}{\partial t}(\theta,s) + \vec{V}\cdot\nabla(\theta,s) + w\frac{\partial}{\partial z}(\theta,s) = (Q,\sigma), \qquad (7)$$

where Q and σ denote source functions. The continuity equation (4)

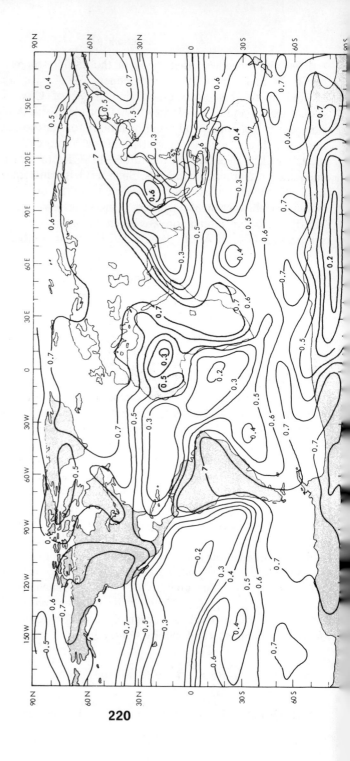

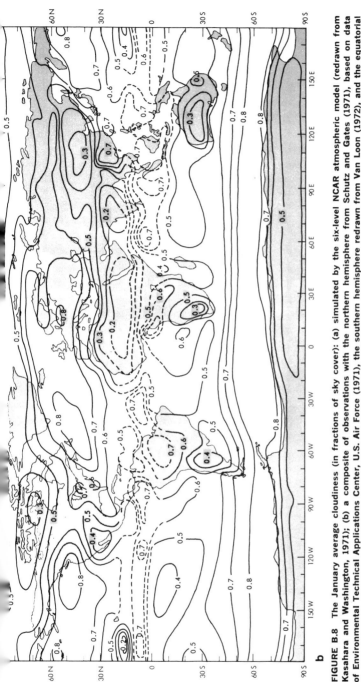

FIGURE B.8 The January average cloudiness (in fractions of sky cover): (a) simulated by the six-level NCAR atmospheric model (redrawn from Kasahara and Washington, 1971); (b) a composite of observations with the northern hemisphere from Schutz and Gates (1971), based on data of Environmental Technical Applications Center, U.S. Air Force (1971), the southern hemisphere redrawn from Van Loon (1972), and the equatorial regions (dashed lines, for December-January-February) redrawn from Kasahara and Washington (1971), based on data of Clapp (1964).

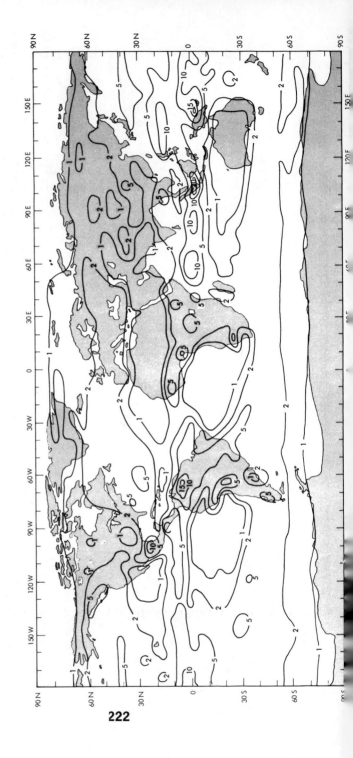

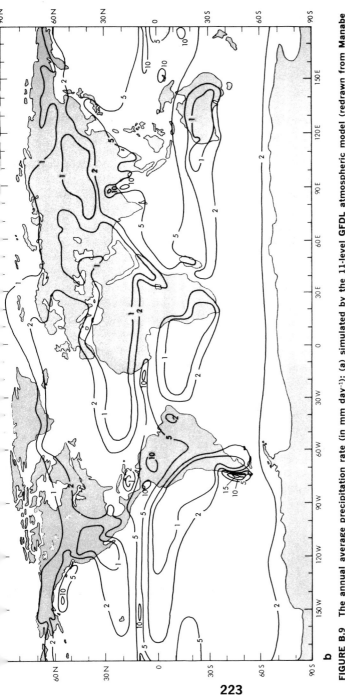

FIGURE B.9 The annual average precipitation rate (in mm day^{-1}): (a) simulated by the 11-level GFDL atmospheric model (redrawn from Manabe et al., 1974b); (b) observed, redrawn from Budyko (1956).

may be simplified by considering the ocean to be incompressible, in which case we may write

$$\frac{\partial w}{\partial z} + \nabla \cdot \vec{V} = 0. \tag{8}$$

The oceanic equation of state may be written symbolically as

$$\rho = \rho(\theta, s, p), \tag{9}$$

where the actual expression is a polynomial of high order, whose coefficients have been determined by laboratory experiments. To close the system, expressions must be chosen for $\vec{F}$ [in the simplified form of Eq. (1)] and for Q and σ in terms of the dependent variables. As in the case of atmospheric models, this closure is an important problem in the formulation of oceanic models and includes the parameterization of the mesoscale oceanic eddies.

Solution Methods

The predictive equations for momentum, temperature, and salinity given in the previous section are generally approximated by centered differences of second-order accuracy, with care taken to conserve both linear and quadratic quantities. The numerical methods that have been used successfully for large-scale models of the atmosphere are usually further modified by the exclusion of external gravity waves from the system. This permits the use of a time step 50 to 100 times larger than is possible for the atmosphere. This is accomplished by requiring the total, vertically integrated flow to be divergence-free, in which case it is possible to specify the total transport by a stream function.

The numerical time integration of an oceanic GCM formulated in this manner proceeds by a combination marching and jury process, involving the explicit prediction of θ, s and $\vec{V}$, and the iterative solution for the total transport stream function. Takano (1974) has recently introduced the implicit treatment of Rossby waves, which allows a considerably longer time step with little loss in accuracy for problems in which the emphasis is on low-frequency oceanic phenomena.

Selected Climatic Simulations

To illustrate the characteristic climatic performance of global oceanic GCM's, we here present comparative solutions from the recent models of Takano et al. (1974), Cox (1974), and Alexander (1974). A number of characteristics of these models are given in Table B.2. These

TABLE B.2 Characteristics of Recent Global Ocean Circulation Models

Feature	UCLA [a]	GFDL [b]	Rand [c]
Number of levels	5	9	2
Horizontal spacing	$\Delta\phi=4°$	$\Delta\phi=2°$	$\Delta\phi=4°$
	$\Delta\lambda=2.5°$	$\Delta\lambda=2°$	$\Delta\lambda=5°$
Salinity	No	Yes	No
Depth	4 km	Actual	300 m
Horizontal mixing [d]	$A_M=10^9$	$A_M=2\times 10^9$	$A_M=7\times 10^9$
($cm^2\ sec^{-1}$)	$A_H=2.5\times 10^7$	$A_H=10^7$	$A_H=5\times 10^7$
Initial condition	Isothermal	Observed T,s	Observed T
Time span of experiment	30 yr	2.5 yr	1.5 yr
Upper boundary condition	Momentum flux, thermal forcing	Momentum flux, T,s specified	Momentum flux, heat flux

[a] Takano et al. (1974); see also Mintz and Arakawa (1974) and Takano (1974).
[b] Cox (1974); see also Bryan et al. (1974).
[c] Alexander (1974).
[d] Here A_M and A_H denote the eddy coefficients for momentum and heat, respectively.

models are currently undergoing further development, and similar oceanic models are under construction at NCAR and at GISS. It is a general characteristic of all such oceanic models that the circulation is dominated by the large values of viscosity, and further efforts are required to extend the solutions into the less viscous and more nonlinear range.

Surface Current

The annual surface current simulated by the nine-level GFDL model (Cox, 1974) is shown in Figure B.10, along with the observed currents for February and March. The February surface currents simulated by the five-level UCLA model (Takano et al., 1974) and the March 1 surface currents simulated by the two-level Rand model (Alexander, 1974) are similarly shown in Figures B.11 and B.12. In each case the overall pattern of the large-scale circulation is simulated successfully, although in general the strength of the equatorial and major western boundary currents is underpredicted. We may note, however, that the UCLA model's solution represents a 30-year integration, the GFDL solution is for 2.5 years, and the Rand solution is for 1.5 years. Closer examination reveals that the simulated surface currents diverge from the equator somewhat more than do those observed, due to the models' effective averaging over the depth of the surface Ekman layer.

FIGURE B.10 The annual average ocean surface current: (a) simulated by the nine-level GFDL oceanic model (Cox, 1974); (b) observed (schematic) for February–March from Bryan et al. (1974), based on data of Sverdrup et al. (1942).

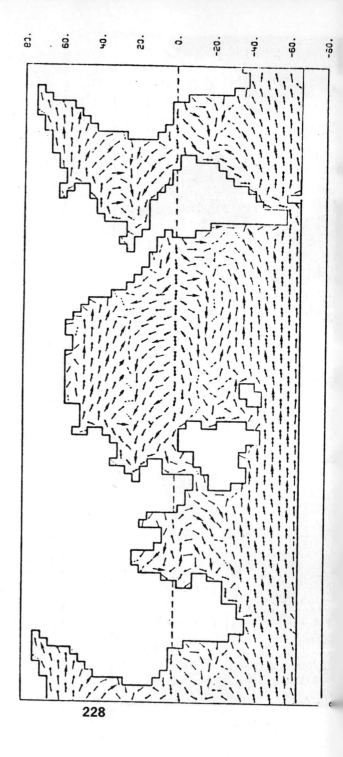

228

FIGURE B.11 The February average ocean surface current: (a) simulated by the five-level UCLA oceanic model (Takano et al., 1974), with half-barbed arrows denoting a speed of 1.3 cm sec^{-1}, and each additional half-barbed denoting a doubling of speed; (b) observed (schematic) for February–March from Bryan et al. (1974), based on data of Sverdrup et al. (1942).

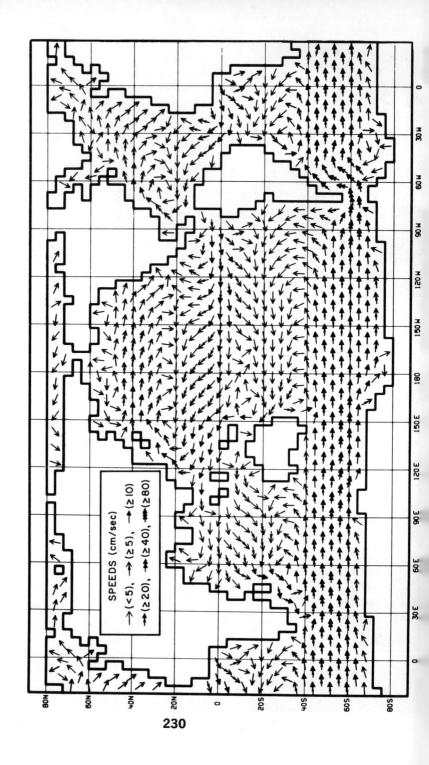

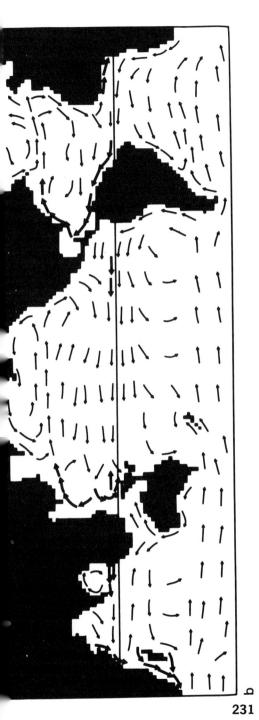

FIGURE B.12 The February–April ocean surface currents: (a) simulated for March 1 by the Rand two-level oceanic model (Alexander, 1974); (b) observed (schematic) for February–March from Bryan et al. (1974), based on data of Sverdrup et al. (1942).

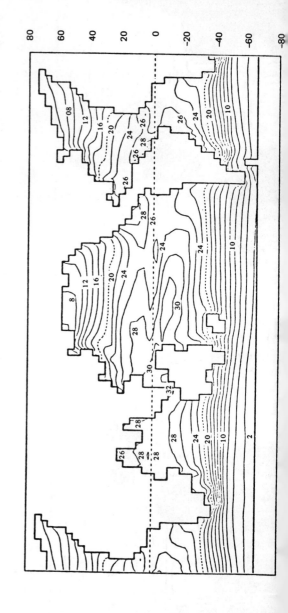

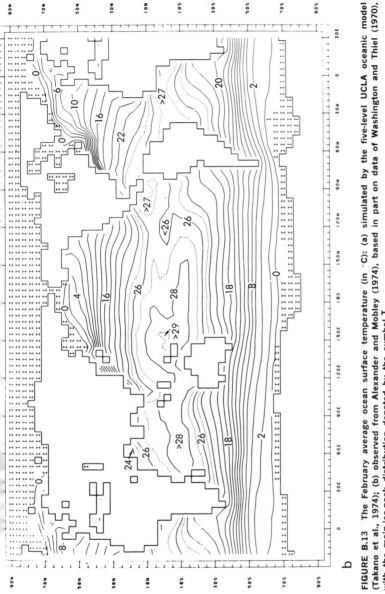

FIGURE B.13 The February average ocean surface temperature (in °C): (a) simulated by the five-level UCLA oceanic model (Takano et al., 1974); (b) observed from Alexander and Mobley (1974), based in part on data of Washington and Thiel (1970), with the main-ice-pack distribution denoted by the symbol I.

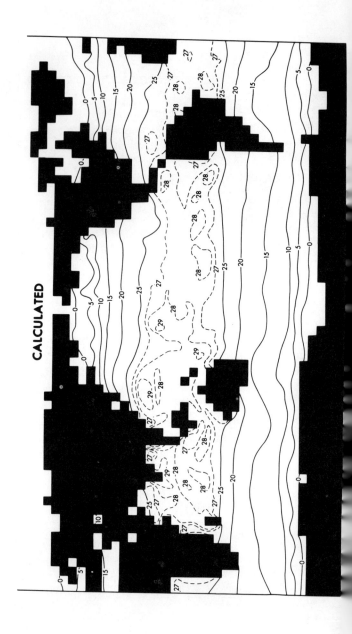

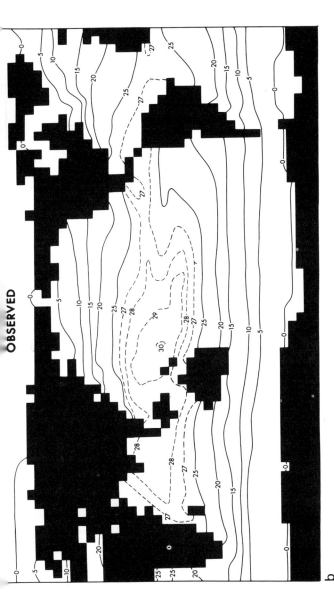

FIGURE B.14 The annual average ocean surface temperature (in °C): (a) simulated by the coupled GFDL model; (b) observed, based on Navy Hydrographic Office data. From Bryan et al. (1974).

Sea-Surface Temperature

The February sea-surface temperature simulated by the five-level UCLA model (Takano et al., 1974) is shown in Figure B.13, along with the observed distribution. To some extent the agreement of the simulation with observation is due to the use of observed components in the surface heat balance condition. The prediction of low surface temperatures at the equator, however, is a feature entirely due to the model's internal dynamics.

Coupled Ocean–Atmosphere Models

As has been previously noted, a dynamical model adequate for the study of climatic variation should include the coupling of the ocean and atmosphere. The first attempt at such coupling was made by Manabe and Bryan (1969) for an idealized ocean basin and later extended by Wetherald and Manabe (1972). In such a joint model, the net fluxes of heat, moisture, and momentum at the air–sea interface are determined by the atmospheric model, while the ocean model in turn provides the sea-surface temperature as a lower boundary condition for the atmosphere.

These studies at GFDL have recently been extended to the entire world ocean, and the results of a coupled numerical integration are now available (Manabe et al., 1974a; Bryan et al., 1974). In this study, the nine-level GFDL atmospheric model was integrated for 0.85 of a year simulated time, while a twelve-layer ocean model was integrated for 256 years' time. The annual sea-surface temperatures simulated in this joint model are shown in Figure B.14, along with the observed distribution. The general level of accuracy may be considered satisfactory, especially in view of the absence of any specification of observed quantities at the air–sea interface. Much further development and testing of such coupled models is required so that their potential for the study of global climatic variations may be realized.

REFERENCES

Alexander, R. C., 1974: Ocean circulation and temperature prediction model, The Rand Corporation, Santa Monica, Calif. (in preparation).

Alexander, R. C., and R. L. Mobley, 1974: Updated global monthly mean ocean surface temperatures, R–1310–ARPA, The Rand Corporation, Santa Monica, Calif. (in preparation).

Arakawa, A., and Y. Mintz, 1974: The UCLA atmospheric general circulation model, Dept. of Meteorol., U. of Calif., Los Angeles, 403 pp.

Bryan, K., and M. D. Cox, 1967: A numerical investigation of the oceanic general circulation, *Tellus, 19*:54–80.
Bryan, K., and M. D. Cox, 1968: A nonlinear model of an ocean driven by wind and differential heating. Parts I and II, *J. Atmos. Sci., 25*:945–978.
Bryan, K., S. Manabe, and R. C. Paconowski, 1974: Global ocean–atmosphere climate model. Part II. The oceanic circulation, Geophysical Fluid Dynamics Laboratory/NOAA, Princeton U., Princeton, N.J., 55 pp. *J. Phys. Oceanog.* (to be published).
Budyko, M. I., 1956: *Heat Balance of the Earth's Surface*, U.S. Weather Bureau, Washington, D.C., 259 pp.
Budyko, M. I., 1969: The effect of solar radiation variations on the climate of the earth, *Tellus, 21*:611–619.
Charney, J. G., and N. A. Phillips, 1953: Numerical integration of the quasigeostrophic equations for barotropic and simple baroclinic flows, *J. Meteorol., 10*: 71–99.
Charney, J. G., R. Fjortoft, and J. von Neumann, 1950: Numerical integration of the barotropic vorticity equation, *Tellus, 2*:237–254.
Clapp, P. F., 1964: Global cloud cover for seasons using TIROS nephanalysis, *Mon. Wea. Rev., 92*:495–507.
Cox, M. D., 1970: A mathematical model of the Indian Ocean, *Deep-Sea Res., 17*:47–75.
Cox, M. D., 1974: A baroclinic numerical model of the world ocean: preliminary results, in *Numerical Models of the Ocean Circulation*, Proceedings of Symposium Held at Durham, New Hampshire, October 17–20, 1972, National Academy of Sciences, Washington, D.C. (in press).
Crutcher, H. L., and J. M. Meserve, 1970: *Selected level heights, temperatures and dew points for the Northern Hemisphere*, NAVAIR 50–1C–52, Naval Weather Service, Washington, D.C., 370 pp.
Environmental Technical Applications Center, U.S. Air Force, 1971: *Northern Hemisphere Cloud Cover*, Project 6168, Washington, D.C. (unpublished data).
Friedrich, H. J., 1970: Preliminary results from a numerical multilayer model for the circulation of the North Atlantic, *Dtsch. Hydrogr. Z., 23*:145–164.
GARP, Joint Organizing Committee, 1974: Modelling for the first GARP global experiment, *GARP Publications Series, No. 14*, World Meteorological Organization, Geneva, 261 pp.
Gates, W. L., 1972: The January global climate simulated by the two-level Mintz-Arakawa model: a comparison with observation, R–1005–ARPA, The Rand Corporation, Santa Monica, Calif., 107 pp. (to be published).
Gilbert, K. D., 1974: A review of numerical models of oceanic general circulation, The Rand Corporation, Santa Monica, Calif. (in preparation).
Haltiner, G. J., 1971: *Numerical Weather Prediction*, Wiley, New York, 317 pp.
Haney, R. L., 1974: A numerical study of the large-scale response of an ocean circulation to surface-heat and momentum flux, *J. Phys. Oceanog.* (to be published).
Heastie, H., and P. M. Stephenson, 1960: Upper winds over the world, Part I, H.M.S.O., London, *Geophys. Mem. No. 103, 13*(3).
Holland, W. R., and A. D. Hirschman, 1972: A numerical calculation of the circulation in the North Atlantic ocean, *J. Phys. Oceanog., 2*:336–354.
Kasahara, A., and W. M. Washington, 1971: General circulation experiments with

a six-layer NCAR model, including orography, cloudiness and surface temperature calculation, *J. Atmos Sci., 28*:657–701.

Lorenz, E. N., 1967: *The Nature and Theory of the General Circulation of the Atmosphere,* World Meteorological Organization, Geneva, 161 pp.

Manabe, S., and K. Bryan, 1969: Climate calculations with a combined ocean–atmosphere model, *J. Atmos. Sci., 26*:786–789.

Manabe, S., K. Bryan, and M. J. Spelman, 1974a: A global atmosphere–ocean climate model. Part 1. The atmospheric circulation, Geophysical Fluid Dynamics Laboratory/NOAA, Princeton U., Princeton, N.J., 76 pp. (to be published).

Manabe, S., D. G. Hahn, and J. L. Holloway, Jr., 1974b: The seasonal variation of the tropical circulation as simulated by a global model of the atmosphere, *J. Atmos. Sci., 31*:43–83.

Mintz, Y., 1965: Very long-term global integration of the primitive equations of atmospheric motion, in WMO-IUGG Symposium on Research and Development Aspects of Long-range Forecasting, *WMO Tech. Note No. 66,* pp. 141–155.

Mintz, Y., and A. Arakawa, 1974: The UCLA oceanic general circulation model, Dept. of Meteorol., U. of Calif., Los Angeles, 138 pp.

Mintz, Y., A. Katayama, and A. Arakawa, 1972: Numerical simulation of the seasonally and inter-annually varying tropospheric circulation, in *Proc. Survey Conf., CIAP,* Department of Transportation, Cambridge, Mass., pp. 194–216.

Ocean Affairs Board, 1974: *Numerical Models of the Ocean Circulation,* Proceedings of Symposium Held at Durham, New Hampshire, October 17–20, 1972, National Academy of Sciences, Washington, D.C. (in press).

Phillips, N. A., 1956: The general circulation of the atmosphere: a numerical experiment, *Q. J. R. Meteorol. Soc., 82*:123–164.

Robinson, G. D., 1971: Review of climate models, in *Man's Impact on the Climate,* W. H. Matthews, W. W. Kellogg, and G. D. Robinson, eds., MIT Press, Cambridge, Mass., pp. 205–215.

Saltzman, B., and A. D. Vernekar, 1971: An equilibrium solution for the axially symmetric component of the earth's macroclimate, *J. Geophys. Res., 76*:1498–1524.

Schneider, S. H., and R. E. Dickinson, 1974: Climate modeling, *Rev. Geophys. Space Phys., 12*:447–493.

Schutz, C., and W. L. Gates, 1971: Global climatic data for surface, 800 mb: January, R–915–ARPA, The Rand Corporation, Santa Monica, Calif., 173 pp.

Sellers, W. D., 1973: A new global climatic model, *J. Atmos. Sci., 12*:241–254.

Smagorinsky, J., 1963: General circulation experiments with the primitive equations: I. The basic experiment, *Mon. Wea. Rev., 91*:99–164.

Smagorinsky, J., 1970: Numerical simulation of the global atmosphere, in *The Global Circulation of the Atmosphere,* G. A. Corby, ed., Royal Meteorol. Soc., London, pp. 24–41.

Smagorinsky, J., 1974: Global atmospheric modeling and the numerical simulation of climate, in *Weather Modification,* W. N. Hess, ed., Wiley, New York (to be published).

Somerville, R. C. J., P. H. Stone, M. Halem, J. E. Hansen, J. S. Hogan, L. M. Druyan, G. Russell, A. A. Lacis, W. J. Quirk, and J. Tenenbaum, 1974: The GISS model of the global atmosphere, *J. Atmos. Sci., 31*:84–117.

Stommel, H., 1948: The westward intensification of wind-driven ocean currents, *Trans. Am. Geophys. Union, 29*:202–206.

Sverdrup, H. U., 1947: Wind-driven currents in a baroclinic ocean; with application to the equatorial currents of the eastern Pacific, *Proc. Nat. Acad. Sci., U.S.,* *33*:318–326.

Sverdrup, H. U., M. W. Johnson, and R. H. Fleming, 1942: *The Oceans,* Prentice-Hall, Englewood Cliffs, N.J., 1087 pp.

Takano, K., 1974: A general circulation model for the world ocean, Dept. of Meteorol., U. of Calif., Los Angeles (unpublished report).

Takano, K., Y. Mintz, and Y. J. Han, 1974: Numerical simulation of the seasonally varying baroclinic world ocean circulation, Dept. of Meteorol., U. of Calif., Los Angeles (unpublished).

Taljaard, J. J., H. Van Loon, H. L. Crutcher, and R. L. Jenne, 1969: *Climate of the upper air: Southern Hemisphere, 1. Temperatures, dew points and heights at selected pressure levels,* NAVAIR 50–1C–55, Naval Weather Service, Washington, D.C.

Van Loon, H., 1972: Cloudiness and precipitation in the Southern Hemisphere, in Meteorology of the Southern Hemisphere, *Meteorol. Monogr., 13,* No. 35, Am. Meteorol. Soc., Boston, Mass., pp. 101–111.

Vonder Haar, T. H., and A. H. Oort, 1973: New estimate of annual poleward energy transport by Northern Hemisphere oceans, *J. Phys. Oceanog., 3*:169–172.

Washington, W. M., and L. G. Thiel, 1970: Digitized global monthly mean ocean surface temperatures, *Tech. Note 54,* National Center for Atmospheric Research, Boulder, Colo.

Wetherald, R. T., and S. Manabe, 1972: Response of the joint ocean–atmosphere model to the seasonal variation of the solar radiation, *Mon. Wea. Rev., 100*: 42–59.

Willson, M. A. G., 1973: Statistical–dynamical modelling of the atmosphere, *Internal Scientific Report No. 17,* Commonwealth Meteorol. Res. Centre, Melbourne, 53 pp.